BestMasters

Mit „BestMasters" zeichnet Springer die besten Masterarbeiten aus, die an renommierten Hochschulen in Deutschland, Österreich und der Schweiz entstanden sind. Die mit Höchstnote ausgezeichneten Arbeiten wurden durch Gutachter zur Veröffentlichung empfohlen und behandeln aktuelle Themen aus unterschiedlichen Fachgebieten der Naturwissenschaften, Psychologie, Technik und Wirtschaftswissenschaften. Die Reihe wendet sich an Praktiker und Wissenschaftler gleichermaßen und soll insbesondere auch Nachwuchswissenschaftlern Orientierung geben.

Springer awards "BestMasters" to the best master's theses which have been completed at renowned Universities in Germany, Austria, and Switzerland. The studies received highest marks and were recommended for publication by supervisors. They address current issues from various fields of research in natural sciences, psychology, technology, and economics. The series addresses practitioners as well as scientists and, in particular, offers guidance for early stage researchers.

More information about this series at http://www.springer.com/series/13198

Paulo Guilherme Santos

Diagonalization in Formal Mathematics

 Springer Spektrum

Paulo Guilherme Santos
Universidade Nova de Lisboa
Lisboa, Portugal

The work of this thesis was presented at FCT-UNL for the Master's Degree in Pure Mathematics. It was funded by the FCT project: PTDC/MHC-FIL/2583/2014.

ISSN 2625-3577   ISSN 2625-3615   (electronic)
BestMasters
ISBN 978-3-658-29110-5   ISBN 978-3-658-29111-2   (eBook)
https://doi.org/10.1007/978-3-658-29111-2

This Springer Spektrum imprint is published by the registered company Springer Fachmedien Wiesbaden GmbH part of Springer Nature.
The registered company address is: Abraham-Lincoln-Str. 46, 65189 Wiesbaden, Germany

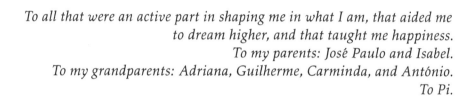

To all that were an active part in shaping me in what I am, that aided me to dream higher, and that taught me happiness.
To my parents: José Paulo and Isabel.
To my grandparents: Adriana, Guilherme, Carminda, and António.
To Pi.

Acknowledgment

I am very thankful to my Advisers—Professor Isabel Oitavem and Professor Reinhard Kahle—for their help in the work developed in the current thesis. I am deeply thankful to Professor Reinhard Kahle for helping me fulfilling my professional dreams. In our meetings—a time where I have learned to be creative and to always be curious—, I have had the opportunity to contact with a great variety of subjects and new ideas that have been essential to the investigation that I have developed under his tuition. I am profoundly grateful to my parents—José Paulo and Isabel—, to my grandparents—Adriana, Guilherme, Carminda, and António—, and to my girlfriend Pilar. I also thank to my friends and family for their love.

The present work started under the Gulbenkian scholarship "Novos Talentos em Matemática" where I had Professor Kahle as my supervisor. The thesis was developed with the collaboration of Tübingen University via Professor Kahle and presented at FCT-NOVA. The work was developed with support of the project *Hilbert's 24th Problem*, funded by the Portuguese Science Foundation FCT (PTDC/MHC-FIL/2583/2014).

Contents

1 Preliminaries

As the main subject of the current thesis is Mathematical Logic, we will assume that the main definitions and results of Logic are known (the definition of formula, the connectives, first-order theories, etc); for more information in introductory notions of Logic see: [Sho18], [Bar93], [Rau06], and [EFT96]. We will also assume the main definitions and results of Category Theory, a domain where we will use the right-to-left notation (in the rest we will use the usual function notation): af will denote, in the context of categories, the composition of f with a (see [Lan13] for more informations).

We continue the preliminaries by remembering the main notions needed to study theories of Arithmetic. Following the notation of [Rau06], let $\mathbf{F}_n$ denote the set of all n-ary functions with arguments and values in $\mathbb{N}$ and let $\mathbf{F} := \bigcup_{n \in \mathbb{N}} \mathbf{F}_n$. For $f \in \mathbf{F}_m$ and $g_1, \ldots, g_m \in \mathbf{F}_n$, we call $h : \vec{a} \mapsto f(g_1(\vec{a}), \ldots, g_m(\vec{a}))$ the (generalised) *composition* of f and g_i and write $h = f[g_1, \ldots, g_m]$. The set of *primitive recursive* functions is the minimal set of function on $\mathbb{N}$ such that:

Initial The constant function equal to 0, the successor function S, and the projection functions $I_v^n : \vec{a} \mapsto a_v$ ($1 \leq v \leq n$, $n \in \mathbb{N}$) are primitive recursive;

Oc If $h \in \mathbf{F}_m$ and $g_1, \ldots, g_m \in \mathbf{F}_n$ are primitive recursive, then $f = h[g_1, \ldots, g_m]$ is primitive recursive;

Op If $g \in \mathbf{F}_n$ and $h \in \mathbf{F}_{n+2}$ are primitive recursive, then so is $f \in \mathbf{F}_{n+1}$ uniquely determined by the equations

$$f(\vec{a}, 0) = g(\vec{a}); \qquad f(\vec{a}, S(b)) = h(\vec{a}, b, f(\vec{a}, b)).$$

The set of *recursive* functions is the minimal set of function on $\mathbb{N}$ that includes the primitive recursive functions and obeys:

© Springer Fachmedien Wiesbaden GmbH, part of Springer Nature 2020
P. G. Santos, *Diagonalization in Formal Mathematics*, BestMasters,
https://doi.org/10.1007/978-3-658-29111-2_1

Oμ If $g \in \mathbf{F}_{n+1}$ is recursive and is such that $\forall \vec{a}.\exists b.g(\vec{a}, b) = 0$, then f given by $f(\vec{a}) = \mu b[g(\vec{a}, b) = 0]$ is also recursive, where the right-hand side denotes the smallest b such that $g(\vec{a}, b) = 0$.

A predicate $P \subseteq \mathbb{N}^n$ is called primitive recursive (respectively, recursive) if the *characteristic function* χ_P of P defined by

$$\chi_P(\vec{a}) = \begin{cases} 1, & \text{in case } P(\vec{a}) \\ 0, & \text{in case } \neg P(\vec{a}) \end{cases}$$

is primitive recursive (respectively, recursive). It is a well-known fact that the predicates | (divides) and prim (being a prime number) are primitive recursive [Rau06, p. 171,172]. We will assume as given the basic facts and definitions about primitive recursive functions and about recursive functions (for example, their relation to computable functions). For more details see [Rau06, p. 165–175] or [Smi13, p. 83–98]. We remember the following primitive recursive functions [Rau06]:

Prime Numeration The unique function given by

$$p_0 := 2; \quad p_{n+1} := \mu q \leqslant p_n! + 1 \, [\mathrm{prim}(q) \wedge q > p_n].$$

Sequence Code We define the *sequence code* of $\langle a_0, \ldots, a_n \rangle$ as being

$$(a_0, \ldots, a_n) := p_0^{a_0+1} \cdots p_n^{a_n+1} = \prod_{i \leqslant n} p_i^{a_i+1}$$

(we will commonly interchange between the notations $(a_0, \ldots, a_n)$ and $\langle a_0, \ldots, a_n \rangle$). It is clear that if $(a_0, \ldots, a_n) = (b_0, \ldots, b_m)$, then $m = n$ and, for all $0 \leq i \leq n$, $a_i = b_i$. We define the primitive recursive predicate Seq by

$$\mathrm{Seq}(a) := a \neq 0 \wedge \forall p \leq a. \forall q \leq p.(\mathrm{prim}(p) \wedge \mathrm{prim}(q) \wedge p|a \rightarrow q|a).$$

Length Function We define the function $a \mapsto \mathrm{ln}(a)$ by:

$$\mathrm{ln}(a) := \mu k \leqslant a [p_k \nmid a].$$

It is clear that $\mathrm{ln}(1) = 0$ and if $a = (a_0, \ldots, a_n)$, then $\mathrm{ln}(a) = n + 1$.

Component-Recognition Function The function $(a, i) \mapsto (a)_i$ is defined by:

$$(a)_i := \mu k \leqslant a\left[p_i^{k+2} \nmid a\right].$$

The intuitive meaning of the previous definition is clear. We also set $(a)_{last} = (a)_{\ln(a) \dot{-} 1}$, where $\dot{-}$ is as defined in [Rau06, p. 219].

Arithmetic Concatenation We define the function $\star$ by

$$a \star b := a \cdot \prod_{i < \ln(b)} p_{\ln(a)+i}^{(b)_i+1}$$

if $\text{Seq}(a)$ and $\text{Seq}(b)$, and $a \star b := 0$ otherwise.

We recall here that Q is the first-order theory of Arithmetic (with a constant 0, with a unary function-symbol S, with two binary function-symbols $+$ and $\times$, see [Smi13, p. 30]) with the following axioms [Smi13, p. 55,56]:

Axiom 1 $\forall x.\ 0 \neq S(x)$;

Axiom 2 $\forall x.\forall y.\ (S(x) = S(y) \rightarrow x = y)$;

Axiom 3 $\forall x.\ (x \neq 0 \rightarrow \exists y.\ x = S(y))$;

Axiom 4 $\forall x.\ x + 0 = x$;

Axiom 5 $\forall x.\forall y.\ x + S(y) = S(x + y)$;

Axiom 6 $\forall x.\ x \times 0 = 0$;

Axiom 7 $\forall x.\forall y.\ x \times S(y) = (x \times y) + x$.

PA, [Smi13, p. 72], is the first-order theory with the same language as Q, with the same axioms of Q with the exception of Axiom 3 (because it turns out that this axiom can be proved from the others in PA), and every sentence that is the universal closure of an instance of

Induction Schema $\varphi(0) \wedge (\forall x.\ (\varphi(x) \rightarrow \varphi(S(x)))) \rightarrow \forall x.\ \varphi(x)$, where $\varphi(x)$ vanish over all formulas of the language of Arithmetic. Given a natural number n we use the common notation $\overline{n}$ to denote the numeral n in a theory of Arithmetic, i.e., to denote $S^n(0)$.

Following [Smi13, p. 99–103], we introduce some important notions about capturing functions in a theory of Arithmetic. The one-place function f is *captured as a function* by $\varphi(x,y)$ in a theory of Arithmetic T just if, for any $m, n \in \mathbb{N}$, if $f(m) = n$, then $\vdash_T \forall y.(\varphi(\overline{m}, y) \leftrightarrow y = \overline{n})$. Furthermore, the one-place function f is *fully captured* as a function by $\varphi(x,y)$ just if $\vdash_T \forall x.\exists! y.\varphi(x,y)$ and for any $m, n \in \mathbb{N}$,

(i) If $f(m) = n$, then $\vdash_T \varphi(\overline{m}, \overline{n})$;

(ii) If $f(m) \neq n$, then $\vdash_T \neg\varphi(\overline{m}, \overline{n})$.

We say that a theory of Arithmetic T is primitive recursive adequate (respectively, strongly primitive recursive adequate) if, for every primitive recursive function f, there is a corresponding formula φ in T that captures it as a function (respectively, fully captures it as a function). It is a well-known fact that Q is primitive recursive adequate and that PA is strongly primitive recursive adequate [Smi13, p. 116]. This means that Q can capture every primitive recursive relation (and PA fully capture). This means that if $R \subseteq \mathbb{N}$ is a primitive recursive relation, then there is a formula φ such that for all $m_1, \ldots, m_n \in \mathbb{N}$, $R(m_1, \ldots, m_n)$ holds if, and only if, $\vdash_{\text{PA}} \varphi(\overline{m_1}, \ldots, \overline{m_n})$; and $R(m_1, \ldots, m_n)$ does not hold if, and only if, $\vdash_{\text{PA}} \neg\varphi(\overline{m_1}, \ldots, \overline{m_n})$ (see [Lin17, p. 8]).

In the remaining of the preliminaries, we are going to consider a fixed first-order theory T that is a consistent primitively recursive axiomatised extension of PA (see [Smi13, p. 148]). Now we move to code string of symbols ξ into natural number $\#\xi$. We fix the following:

s	$=$	$\neg$	$\wedge$	$\vee$	$($	$)$	0	S	$+$	$\times$	v_0	v_1	$\cdots$
$\#s$	1	3	5	7	9	11	13	15	17	19	21	23	$\cdots$

Given a string of symbols $\xi = s_0 \cdots s_n$, its *Gödel number*, $\#\xi$, is given by:

$$\#\xi := (\#s_0, \ldots, \#s_n) = p_0^{1+\#s_0} \cdots p_n^{1+\#s_n}.$$

It is clear that given strings of symbols ξ and η, then

$$\#(\xi\eta) = \#\xi \star \#\eta.$$

Given a string of symbols ξ, we denote $\overline{\#\xi}$ simply by $\ulcorner \xi \urcorner$. In particular, if φ is a formula, $\ulcorner \varphi \urcorner$ denotes $\overline{\#\varphi}$. Let us consider the primitive recursive functions $num(n)$ defined by:

$$num(0) := \#0; \qquad num(x+1) := \#S \star num(n).$$

We have that num is fully captured by a function-symbol num. One can easily construct (see [Bar93, p. 837]) a primitive recursive function sub such that if $\varphi(v_i)$ is a formula where v_i is free, $r(v_i)$ is a term where v_i is free, and t is a term, then

$$sub(\#\varphi(v_i), i, \#t) = \#\varphi(t); \qquad sub(\#r(v_i), i, \#t) = \#r(t).$$

Using sub, one can define the following primitive recursive functions:

$$s'(x,y) := \begin{cases} sub(x,i,y) & \text{if } i = \mu j < x \Big[v_j \text{ occurs free in } x \Big] \\ x & \text{if } i \text{ does not exist.} \end{cases}$$

and

$$s(x,y) := s'(x, num(y)).$$

As s and s' are a primitive recursive functions and T is an extension of PA, then s can be fully captured by a function-symbol s and s' by a function-symbol s'. We usually abbreviate (confirm [Bar93, p. 837]) s($\ulcorner \varphi(x) \urcorner, y$) by $\ulcorner \varphi(\dot{y}) \urcorner$, if $\varphi(x)$ is a formula, and s($\ulcorner t(x) \urcorner, y$) by $\ulcorner t(\dot{y}) \urcorner$, if $t(x)$ is a term. Thus, if x is the only free-variable of the formula $\varphi(x)$, if x is the only free-variable of the term $r(x)$, and if t is a closed-term denoting $n \in \mathbb{N}$, then $\vdash_T s(\ulcorner\varphi(x)\urcorner, t) = \ulcorner\psi\urcorner$, where $\vdash_T \psi \leftrightarrow \varphi(\overline{n})$ and $\vdash_T s(\ulcorner r(x)\urcorner, t) = \ulcorner j\urcorner$, where $\vdash_T j = r(\overline{n})$.

Let us consider the following primitive recursive predicates in $\mathbb{N}$ (see [Bar93, p. 836–837] or [Rau06, p. 231–234]):

Being the Gödel Number of a Term We define the predicate *Term* by: *Term*(n) holds if, and only if, n is the Gödel number of a term.

Being the Gödel Number of a Formula We will consider the predicate *Form* defined by: *Form*(n) is true if, and only if, n is the Gödel number of a formula.

Being the Gödel Number of an Axiom of T We define the predicate *Axiom* by: *Axiom*(n) holds if, and only if, n is the Gödel number of an axiom.

As T is primitively recursive axiomatised extension of PA, there are formulas in T—Term, Form, and Axiom—that capture the previous predicates in $\mathbb{N}$. We define the primitive recursive functions $\tilde{\neg}, \tilde{\wedge}, \tilde{\rightarrow}$ as follows:

$$\tilde{\neg}a := \#\neg \star a; \quad a \tilde{\wedge} b := \#(\star a \star \# \wedge \star b \star \#); \quad a \tilde{\rightarrow} b := \tilde{\neg}(a \tilde{\wedge} \tilde{\neg}b).$$

In the previous definition the parenthesis are occurring both as a syntactic symbol and as a symbol in the meta-language. Following [Rau06, p. 229], let us consider the primitive recursive relations $proof_T$ and bew_T in $\mathbb{N}$ given, respectively, by:

$$proof_T(b) := Seq(b) \wedge b \neq 1 \wedge$$
$$\forall k < \ln(b). \, (Axiom((b)_k) \vee \exists i, j < k. \, (b)_i = (b)_j \tilde{\rightarrow} (b)_k)$$

and by:

$$bew_T(b, a) := proof_T(b) \wedge a = (b)_{last}.$$

The intuitive meaning of $bew_T(b, a)$ is straightforward: b is the Gödel number of a proof (that is a sequence of inferences) of a. We have that there is a formula bew_T in T that captures bew_T. Let us now define the *provability predicate* in T, $\mathcal{P}_T$, by:

$$\mathcal{P}_T(x) := \exists y.\ bew_T(y,x).$$

We have that the Hilbert-Bernays-Löb derivability conditions hold [Smi13, p. 223], i.e., given formulas φ and ψ:

C1.) If $\vdash_T \varphi$, then $\vdash_T \mathcal{P}_T(\ulcorner \varphi \urcorner)$ (the other implication does not necessarily hold);

C2.) $\vdash_T \mathcal{P}_T(\ulcorner \varphi \rightarrow \psi \urcorner) \rightarrow (\mathcal{P}_T(\ulcorner \varphi \urcorner) \rightarrow \mathcal{P}_T(\ulcorner \psi \urcorner))$;

C3.) $\vdash_T \mathcal{P}_T(\ulcorner \varphi \urcorner) \rightarrow \mathcal{P}_T(\ulcorner \mathcal{P}_T(\ulcorner \varphi \urcorner) \urcorner)$.

We now move to construct a very important sentence in T following [Smi13, p. 130–139]. Let us consider the primitive recursive function *diag* defined by:

$$diag(n) = \#(\exists y(y =) \star num(n) \star \# \wedge \star n \star \#).$$

Given a formula φ, $diag(\#\varphi)$ is $\#(\exists y(y = \ulcorner \varphi \urcorner \wedge \varphi))$. Let $Gdl_T(m,n)$ be the primitive recursive relation given by

$$Gdl_T(m,n) := bew_T(m, diag(n)).$$

Let us consider $Gdl_T(x,y)$ as being a formula that captures Gdl_T (see [Smi13, p. 139]). Let us, furthermore, consider

$$U_T(y) := \forall x.\ \neg Gdl_T(x,y),$$

and the *Gödel sentence (in T)*

$$\mathcal{G}_T := \exists y.\ (y = \ulcorner U_T(y) \urcorner \wedge U_T(y)).$$

It is clear that $\vdash_T \mathcal{G}_T \leftrightarrow U_T(\ulcorner U_T(y) \urcorner)$, and thus $\vdash_T \mathcal{G}_T \leftrightarrow \forall x.\ \neg Gdl_T(x, \ulcorner U_T(y) \urcorner)$. It is easy to see [Smi13, p. 139,169] that

$$\vdash_T \mathcal{G}_T \leftrightarrow \neg \mathcal{P}_T(\ulcorner \mathcal{G}_T \urcorner).$$

We end the preliminaries by proving that $\vdash_T \mathcal{P}_T(\ulcorner \varphi \wedge \psi \urcorner) \leftrightarrow \mathcal{P}_T(\ulcorner \varphi \urcorner)$ $\wedge \, \mathcal{P}_T(\ulcorner \psi \urcorner)$, a fact that is going to be useful:

1	$\varphi \wedge \psi \to \varphi$	Logic
2	$\varphi \wedge \psi \to \psi$	Logic
3	$\mathcal{P}_T(\ulcorner \varphi \wedge \psi \to \varphi \urcorner)$	C1.) (1)
4	$\mathcal{P}_T(\ulcorner \varphi \wedge \psi \to \psi \urcorner)$	C1.) (2)
5	$\mathcal{P}_T(\ulcorner \varphi \wedge \psi \to \varphi \urcorner) \to (\mathcal{P}_T(\ulcorner \varphi \wedge \psi \urcorner) \to \mathcal{P}_T(\ulcorner \varphi \urcorner))$	C2.)
6	$\mathcal{P}_T(\ulcorner \varphi \wedge \psi \to \psi \urcorner) \to (\mathcal{P}_T(\ulcorner \varphi \wedge \psi \urcorner) \to \mathcal{P}_T(\ulcorner \psi \urcorner))$	C2.)
7	$\mathcal{P}_T(\ulcorner \varphi \wedge \psi \urcorner) \to \mathcal{P}_T(\ulcorner \varphi \urcorner)$	(3,5)
8	$\mathcal{P}_T(\ulcorner \varphi \wedge \psi \urcorner) \to \mathcal{P}_T(\ulcorner \psi \urcorner)$	(4,6)
9	$\mathcal{P}_T(\ulcorner \varphi \wedge \psi \urcorner) \to \mathcal{P}_T(\ulcorner \varphi \urcorner) \wedge \mathcal{P}_T(\ulcorner \psi \urcorner)$	(7,8)
10	$\varphi \to (\psi \to \varphi \wedge \psi)$	Logic
11	$\mathcal{P}_T(\ulcorner \varphi \to (\psi \to \varphi \wedge \psi) \urcorner)$	C1.) (10)
12	$\mathcal{P}_T(\ulcorner \varphi \to (\psi \to \varphi \wedge \psi) \urcorner) \to (\mathcal{P}_T(\ulcorner \varphi \urcorner) \to \mathcal{P}_T(\ulcorner \psi \to \varphi \wedge \psi \urcorner))$	C2.)
13	$\mathcal{P}_T(\ulcorner \varphi \urcorner) \to \mathcal{P}_T(\ulcorner \psi \to \varphi \wedge \psi \urcorner)$	(11,12)
14	$\mathcal{P}_T(\ulcorner \psi \to \varphi \wedge \psi \urcorner) \to (\mathcal{P}_T(\ulcorner \psi \urcorner) \to \mathcal{P}_T(\ulcorner \varphi \wedge \psi \urcorner))$	C2.)
15	$\mathcal{P}_T(\ulcorner \varphi \urcorner) \wedge \mathcal{P}_T(\ulcorner \psi \urcorner) \to \mathcal{P}_T(\ulcorner \varphi \wedge \psi \urcorner)$	Logic (13,14)
16	$\mathcal{P}_T(\ulcorner \varphi \wedge \psi \urcorner) \leftrightarrow \mathcal{P}_T(\ulcorner \varphi \urcorner) \wedge \mathcal{P}_T(\ulcorner \psi \urcorner)$	(9,15)

2 Introduction

In the context of Mathematics, diagonalization is a very broad term that is linked to several phenomena that vanish from paradoxes to fixed point theorems. Nevertheless, although diagonalization is used to refer to a great variety of phenomena, there is a transversal idea to the use of the term: given a relation $R(x_0,\ldots,x_n)$, it is used to refer to the possibility of $R(a,\ldots,a)$ being the case for a given element a. It is now immediate how fixed point theorems are a case of diagonalization: given a function f, to say that f has a fixed point is equivalent to say that the relation $R(x,y)$ defined by $f(x) = y$ is diagonalisable (has a fixed point). The aim of the present thesis is to study diagonalization in formal systems of Mathematics from self-reference, that is a particular case of diagonalization, to diagonalization in everyday Mathematics.

Before any further discussion on the subject, let us understand the term diagonalization. If one follows a strict way of thinking, diagonalization, in a rough way, refers to the possibility of drawing diagonal lines. This claim is not far from the connotation that we are going to use. If one consider the plot of the function $f(x) = x$ where the image of the function is represented in the y-axis, one gets a perfect diagonal line. This line corresponds to the very simple fact that $y = x$. So, keeping in mind this naïve view of diagonalization as being $y = x$, if one has a relation $R(x,y)$, it make sense that the diagonalization of that relation is simply the operation of replacing the occurrence of y by an x, i.e., $R(x,x)$. As we are generalising fixed point theorems, when speaking about diagonalization of a relation $R(x_0,\ldots,x_n)$, rather than making a universal claims, we make an existential: there is an a such that $R(a,\ldots,a)$ is the case. It is now clear

© Springer Fachmedien Wiesbaden GmbH, part of Springer Nature 2020
P. G. Santos, *Diagonalization in Formal Mathematics*, BestMasters,
https://doi.org/10.1007/978-3-658-29111-2_2

the Mathematical use of the term. For a compilation of important fixed point theorems we recommend [Sha16].

There are two very famous diagonalization arguments due to Cantor [Mos06, p. 10–15]. We start by stating and proving the first result.

Theorem 1. *The set of infinite binary sequences* Δ *is uncountable* ($\Delta :=$ $\{\langle a_0, a_1, \ldots, \rangle | \forall i. a_i = 0 \vee a_i = 1\}$).

Proof. Following [Mos06, p. 10], let us suppose, aiming a contradiction, that there is an enumeration $\Delta = \{\alpha_0, \alpha_1, \ldots\}$ where, for each $n \in \mathbb{N}$, $\alpha_n = \langle a_0^n, a_1^n, \ldots \rangle$ is a sequence of 0's and 1's. Let us now define a new sequence β by:

$$\beta(n) := 1 - a_n^n.$$

We have that $\beta \in \Delta$ and $\beta(n) = 1 - a_n^n \neq a_n^n$. Thus, for each $n \in \mathbb{N}$, $\beta \neq \alpha_n$, which is a contradiction. ∎

The proof of the previous result used a diagonalization of a_n^m that corresponds to a_n^n. More schematically, the proof considered the following diagonal line:

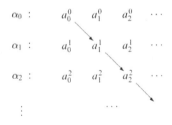

Fig. 2.1: Cantor's argument from [Mos06, p. 11].

Now we present and prove the second result due to Cantor according to [Mos06, p. 14].

Theorem 2. *For every set* A, $A \lneqq \wp(A)$.

Proof. That $A \leq \wp(A)$ is the case, follows from the fact that the function $x \mapsto \{x\}$ is injective.

To complete the proof, let us suppose, aiming a contradiction, that there is a surjective function $\pi : A \to \wp(A)$. Let us consider the set $B = \{x \in A | x \notin \pi(x)\}$. So, for all $x \in A$, $x \in B \leftrightarrow x \notin \pi(x)$. As $B \in \wp(A)$ and as π is surjective, then there is $b \in A$ such that $\pi(b) = B$. But then we get $b \in B \leftrightarrow b \notin B$, which is a contradiction.

∎

In the previous argument there was also a diagonalization: there was a diagonalization of the relation $x \in \pi(y)$. There is a deep relation between the previous proof and Russell's Paradox, a paradox that will be studied in Chapter 5. More examples of diagonalization in Mathematics are going to be studied in Chapter 7.

The use of the term diagonalization is not confined to everyday Mathematics: it is also used in Logic and in Philosophy. As we will see in the present thesis, not only important results of Mathematics are statements about the diagonalization of a relation, but also deep results in Logic are a consequence of this phenomenon. In the context of Logic, diagonalization can have the meaning that we have presented so far, but can also have an adapted version: given a theory T and a formula $\varphi(x_0, \ldots, x_n)$, speaking about the diagonalization of $\varphi(x_0, \ldots, x_n)$ is the same as speaking about the possibility of $\vdash_T \exists x.\ \varphi(x, \ldots, x)$ being the case. A very famous result in Logic that arises from diagonalization is the Diagonalization Lemma that will be studied in detail in Chapter 3. One textbook example of a sentence that arises from the use of a diagonalization is the sentence $\mathcal{G}_T$ that was constructed in Chapter 1. From $\mathcal{G}_T$ and the fact that $\vdash_T \mathcal{G}_T \leftrightarrow \neg \mathcal{P}_T(\ulcorner \mathcal{G}_T \urcorner)$ one can prove one of the most stunning results in Logic, Gödel's First Incompleteness Theorem.

Gödel's First Incompleteness Theorem (adapted from [Bar93]). *If T is a consistent primitively recursive axiomatised extension of Q, then there is a sentence φ such that:*

1.) $\nvdash_T \varphi$;

2.) If $\vdash_T \mathcal{P}_T(\ulcorner\varphi\urcorner) \Longrightarrow \vdash_T \varphi$, then $\nvdash_T \neg\varphi$.

Proof. We will prove that the sentence $\mathcal{G}_T$ satisfies the desired properties. We have that

$$\vdash_T \mathcal{G}_T \leftrightarrow \neg\mathcal{P}_T(\ulcorner\mathcal{G}_T\urcorner). \tag{I}$$

If $\vdash_T \mathcal{G}_T$, then $\vdash_T \mathcal{P}_T(\ulcorner\mathcal{G}_T\urcorner)$ and, by (I), $\vdash_T \neg\mathcal{P}_T(\ulcorner\mathcal{G}_T\urcorner)$. Thus, if $\vdash_T \mathcal{G}_T$, then $\vdash_T \bot$, which is a contradiction. So, $\nvdash_T \varphi$.

Let us suppose that $\vdash_T \mathcal{P}_T(\ulcorner\mathcal{G}_T\urcorner) \Longrightarrow \vdash_T \mathcal{G}_T$. If, $\vdash_T \neg\mathcal{G}_T$, then by (I) we conclude that $\vdash_T \mathcal{P}_T(\ulcorner\mathcal{G}_T\urcorner)$, and by hypothesis follows that $\vdash_T \mathcal{G}_T$; thus, $\vdash_T \bot$, which is a contradiction.

∎

The antecedent condition of 2.) corresponds to the ω-consistency which was assumed by Gödel in his original paper. For more informations on the previous result see: [Rau06], [Smi13], and [Bar93]. We present, without further observations, the definition of ω-consistency ([Smi13, p. 143]): an arithmetical theory T is ω-*inconsistent* if, for some formula $\varphi(x)$, $\vdash_T \varphi(\bar{n})$ holds for all $n \in \mathbb{N}$ and $\vdash_T \neg\forall x.\ \varphi(x)$; T is ω-*consistent* if it is not ω-inconsistent.

There is a very important logical notion that is connected to the notion of diagonalization—self-reference. Self-reference, in logic, refers to the possibility of a given formula speak about itself. Two examples of famous self-referential sentences are the Liar [BGR16], a paradox that is going to be studied in Chapter 5, and Gödel's sentence. The former can be identified with the phrase in natural language "I am false" and with the formal sentence $L \leftrightarrow \neg L$ or, in the presence of a truth predicate, with the sentence $L \leftrightarrow \neg T(L)$, where $T(\cdot)$ denotes the truth predicate. The latter, is the sentence presented in Chapter 1 that satisfies $\vdash_{\text{PA}} \mathcal{G} \leftrightarrow \neg\mathcal{P}_{\text{PA}}(\ulcorner\mathcal{G}\urcorner)$. Intuitively, in the natural language, Gödel's sentence expresses "I am not provable in PA". It is important to observe that in the first example we arrive at a contradiction and in the second example we do not arrive at a contradiction. This means that self-reference can lead to contradictions or simply to unexpected results.

It is clear that in the two previous examples the considered sentences somehow express some non-trivial information about themselves. This property of expressing something about themselves goes beyond the logical aspect of the considered sentences. To see this, just think about the formula $S \leftrightarrow S$. This formula is trivially true and can have any construction, not necessarily a self-referential one. So, it expresses something that we recognise as being trivial, hence it has no non-trivial information about itself. Being in the presence of diagonalization does not guaranty that one is also in the presence of self-reference—for the latter we need the intension of the formula to speak about itself and we need some construction.

This means that self-reference is a property that is hard (or even impossible) to define or study logically and that is characterised by its use and not by a closed definition. As, for the presented reasons, we are considering self-reference as a property that obeys to the previous conditions, we cannot study self-reference using only syntactical means, nevertheless we can recognise indispensable properties that a formula should have to be considered self-referential. Inspired by the previous analysis, we identified two main features that a self-referential formula should have:

1.) A logical layout (or interpretation) of the form $X_0 \leftrightarrow F(X_0)$, where $F(X)$ is a formula that depends on the parameter X that, by its turn, is a formula;

2.) Non-trivial information about itself.

It is important to observe that whenever one can find a formulation of a sentence in the previous conditions that very sentence should be considered self-referential. This does not mean that every formulation of the sentence is self-referential. To see this, the following formulation of the Liar[1] is not self-referential (it does not

---

[1] Some readers might not subscribe to the thesis that this two line paradox is the Liar. Nevertheless, this formulation was presented just to give an example; we are not going to need it for what follows.

satisfy the first feature of self-reference), nevertheless the former formulation of the Liar is self-referential:

L1.) "L2.) is false";

L2.) "L1.) is true".

So, to argue that some sentence is self-referential one needs to find a self-referential formulation of it, but one does not need to show that every formulation of it is self-referential (this last feature probably will never be the case).

The main goal of the current work is to study diagonalization in formal systems with a view towards everyday Mathematics. We will start by studying the Diagonalization Lemma in Chapter 3. After that, in Chapter 4, we will argue that Yablo's Paradox is self-referential. From that, a detailed study of paradoxes and Löb's Theorem will be done in Chapter 5. Chapter 6 will serve to present a general theory of diagonalization, a theory that will be used in Chapter 7 to trace several important results in Mathematics to a common origin: the General Diagonalization Theorem.

3 Two uses of the Diagonalization Lemma

In this chapter, we will use the Diagonalization Lemma for two purposes: to present natural properties related to self-reference that are not decidable, and to argue that one cannot prove the Strong Diagonalization Lemma using the Diagonalization Lemma, i.e., that diagonalization of term is substantially different from diagonalization of formulas.

The Diagonalization Lemma [Rau06, p. 250] is a very famous result in Logic that is, since its birth, deeply linked to Gödel's First Incompleteness Theorem (see [Raa18] for further information). This result states that, fixed a theory T that must be a primitively recursive axiomatised extension of Q, given a one-free-variable-formula $\varphi(x)$, one can always construct a sentence ψ such that $\vdash_T \psi \leftrightarrow \varphi(\ulcorner \psi \urcorner)$. The name of the result is straightforward from the previous equivalence: it states that one can always diagonalize a one-free-variable-formula of T. The sentences that arise from the use of the Diagonalization Lemma are commonly accepted to be self-referential (the sentence ψ that emerges from the use of the Diagonal Lemma applied to $\varphi(x)$ and that satisfies $\vdash_T \psi \leftrightarrow \varphi(\ulcorner \psi \urcorner)$ is commonly accepted to be a sentence that by its very construction expresses, in an intuitive way, "I have the property $\varphi(x)$"). One very natural question in this context is whether one can define formally what self-reference is and whether there are natural properties related to self-reference that are not decidable.

The Strong Diagonalization Lemma, also known as Jeroslow's Lemma (see [Jer73], [Smi13, p. 237]), claims that, for a theory T that is a primitively recursive axiomatised extension of PA and that has a function-symbol for each primitive recursive function, and for

© Springer Fachmedien Wiesbaden GmbH, part of Springer Nature 2020
P. G. Santos, *Diagonalization in Formal Mathematics*, BestMasters,
https://doi.org/10.1007/978-3-658-29111-2_3

each one-free-variable-formula $\varphi(x)$, there is a closed-term t such that $\vdash_T t = \ulcorner\varphi(t)\urcorner$.

It is clear that from the Strong Diagonalization Lemma one can conclude the Diagonalization Lemma: one considers ψ as being $\varphi(t)$ (where t is the closed-term from the Strong Diagonalization Lemma) and one has the following:

1	$\vdash_T t = \ulcorner\varphi(t)\urcorner$	Strong Diagonalization Lemma
2	$\vdash_T \psi \leftrightarrow \varphi(t)$	definition of ψ
3	$\vdash_T \psi \leftrightarrow \varphi(\ulcorner\varphi(t)\urcorner)$	(1)
4	$\vdash_T \psi \leftrightarrow \varphi(\ulcorner\psi\urcorner)$	definition of ψ

The Strong Diagonalization Lemma and the Diagonalization Lemma are really similar in the layout (while one states one equality the other states an equivalence). As was observed before, from the Strong Diagonalization Lemma one can prove the Diagonalization Lemma, that is, reasonings about equivalence of formulas can be captured by reasonings about equality of terms. One very natural question is whether the converse also holds, i.e., can reasonings about equality of terms be captured by reasonings about equivalence of formulas? More generally, is diagonalization of terms substantially different from diagonalization of formulas?

In what follows, we are going to consider T as being a fixed first-order theory that is a consistent primitively recursive axiomatised extension of PA that has a function-symbol to capture each primitive recursive function. In particular we are going to consider a binary function-symbol $*$ that captures the usual concatenation function $\star$ (see the Preliminaries, [Rau06, p. 224], or [Smi13, p. 132]), an unary function-symbol $\neg$ that captures the primitive recursive function $\tilde{\neg}$ (see the Perliminaries, [Rau06, p. 229], or [Smi13, p. 177]) defined by $\tilde{\neg}(a) := (\#\neg) \star a$. Given a formula φ, one has that $\vdash_T \ulcorner\neg\varphi\urcorner = \neg\ulcorner\varphi\urcorner$. We say that a predicate (formula) $\varphi(x)$ is decidable in T [Und16] if, for every $n \in \mathbb{N}$, $\vdash_T \varphi(\overline{n})$ or $\vdash_T \neg\varphi(\overline{n})$.

3.1 Formal Limitations of Predicates Related to Self-Reference

The aim of this section is to argue that important natural properties related to self-reference are not decidable in the theory T. We start by presenting a some-how expected result: not all forms of diagonalization in T are provable, more precisely, there is no way to decide in T if, given a two-free-variable-formula $\varphi(x,y)$, $\exists x.\varphi(x,x)$ holds in T or not.

Theorem 3. *Assume that T is ω-consistent. Then, there is a two-free-variable-formula $\varphi(x,y)$ such that $\exists x.\ \varphi(x,x)$ is independent of T, i.e., $\not\vdash_T \exists x.\ \varphi(x,x)$ and $\not\vdash_T \neg \exists x.\ \varphi(x,x)$.*

Proof. Let us consider $\varphi(x,y)$ as being $x = \ulcorner \mathcal{G}_T \urcorner \land \mathcal{P}_T(y)$. If $\vdash_T \exists x.\ \varphi(x,x)$ was the case, then $\vdash_T \mathcal{P}_T(\ulcorner \mathcal{G}_T \urcorner)$, which is a contradiction. Hence, $\not\vdash_T \exists x.\ \varphi(x,x)$. Let us suppose aiming a contradiction that $\vdash_T \neg \exists x.\ \varphi(x,x)$. Thus, $\vdash_T \forall x.\ \neg\varphi(x,x)$, consequently $\vdash_T \forall x.\ \neg(x = \ulcorner \mathcal{G}_T \urcorner) \lor \neg\mathcal{P}_T(x)$. In particular, $\vdash_T \neg(\ulcorner \mathcal{G}_T \urcorner = \ulcorner \mathcal{G}_T \urcorner) \lor \neg\mathcal{P}_T(\ulcorner \mathcal{G}_T \urcorner)$. As $\vdash_T \ulcorner \mathcal{G}_T \urcorner = \ulcorner \mathcal{G}_T \urcorner$ it follows that $\vdash_T \neg\mathcal{P}_T(\ulcorner \mathcal{G}_T \urcorner)$, thus $\vdash_T \mathcal{G}_T$, which is a contradiction. So, $\not\vdash_T \neg \exists x.\ \varphi(x,x)$.

∎

From the previous result we conclude that the problem of diagonalization in general is not decidable in T. Nevertheless, we can construct very interesting diagonalizations inside T, such as the Diagonalization Lemma.

Diagonalization Lemma [Rau06, p. 250]. *For every one-free-variable-formula $\varphi(x)$ there is a sentence ψ such that*

$$\vdash_T \psi \leftrightarrow \varphi(\ulcorner \psi \urcorner).$$

Proof. Let $f : \mathbb{N} \to \mathbb{N}$ be the function such that for every one-free-variable-formula $\theta(x)$,

$$f(\#\theta(x)) := \#\theta(\ulcorner \theta(x) \urcorner),$$

and if n is not the code of a one-free-variable-formula, then $f(n) :=$ 0. It is clear that f is primitive recursive, so there is a predicate $\text{Diag}(x, y)$ such that

$$\vdash_T \forall y.\, (\text{Diag}(\ulcorner \theta(x) \urcorner, y) \leftrightarrow y = \overline{f(\#\theta(x))}). \tag{I}$$

Let $\beta(z)$ be the formula $\forall y.\, (\text{Diag}(z, y) \to \varphi(y))$. Given a one-free-variable-formula $\theta(x)$ we have that

1	$\vdash_T \beta(\ulcorner \theta(x) \urcorner) \leftrightarrow \forall y.\, (y = \overline{f(\#\theta(x))} \to \varphi(y))$	(I)
2	$\vdash_T \beta(\ulcorner \theta(x) \urcorner) \leftrightarrow \varphi(\overline{f(\#\theta(x))})$	(1)
3	$\vdash_T \beta(\ulcorner \theta(x) \urcorner) \leftrightarrow \varphi(\ulcorner \theta(\ulcorner \theta(x) \urcorner) \urcorner)$	def. of f

In particular, taking $\theta(x)$ as being $\beta(z)$, if follows that

$$\vdash_T \beta(\ulcorner \beta(z) \urcorner) \leftrightarrow \varphi(\ulcorner \beta(\ulcorner \beta(z) \urcorner) \urcorner).$$

The conclusion follows by taking ψ as being $\beta(\ulcorner \beta(z) \urcorner)$.

We say that ψ results from the use of the Diagonalization Lemma if ψ is $\beta(\ulcorner \beta(z) \urcorner)$ for some one-free-variable-formula $\varphi(x)$ (according to the notations introduced in the previous proof). The sentences that arise by the Diagonalization Lemma are commonly accepted to be self-referential. They are so not only because they satisfy $\vdash_T$ $\psi \leftrightarrow \varphi(\ulcorner \psi \urcorner)$, but also by their very construction (using $\beta(z)$). It is important to observe that we are considering in the set-up of the Diagonalization Lemma the sentence ψ as being self-referential and not the equivalence $\psi \leftrightarrow \varphi(\ulcorner \psi \urcorner)$. Being in the presence of a sentence that satisfies the equivalence does not necessarily mean that one is in presence of a self-referential sentence—it is important to have some construction where the intension of the sentence speak about itself is placed, just like the construction of $\beta(\ulcorner \beta(z) \urcorner)$ (for further reading on this see [HV14] and [Pic18]).

There is a version of the Diagonalization Lemma for a two-free-variable-formula that we are not going to prove because the proof is really similar to the one of the Diagonalization Lemma.

Two-Variable Diagonalization Lemma [AM99, p. 208]. *For every two-free-variable-formula $\alpha(x,y)$ there is a one-free-variable-formula $\varphi(x)$ such that*

$$\vdash_T \forall z.\ \varphi(z) \leftrightarrow \alpha(\ulcorner\varphi(x)\urcorner, z).$$

Now we move to establish some relations between the Diagonalization Lemma and self-reference. For that, we introduce the following notion:

Definition 1. We say that a one-free-variable formula $\alpha(x)$ is a *general formula* if, for every sentence ψ that results from the application of the Diagonalization Lemma, $\vdash_T \alpha(\ulcorner\psi\urcorner)$ holds.

It is clear that $\text{Sent}(x)$ and $\text{Form}(x)$ are general formulas (where $\text{Sent}(x)$ is the predicate that captures the sentences of T, and $\text{Form}(x)$ is the predicate that captures formulas of T). If one defines $\delta(x)$ as being the predicate that identifies the sentences that arise by the use of the Diagonalization Lemma (this clearly can be captured by[1] T), then $\delta(x)$ is a general formula. More importantly, if $\sigma(x)$ is a predicate that identifies all self-referential formulas (see [Pic18] for a possible definition), as all sentences that arise from the Diagonalization Lemma are self-referential, it follows that $\sigma(x)$ is a general formula. In fact, by what was previously said, if there is such a predicate $\sigma(x)$, then one should have $\vdash_T \forall x.\ \delta(x) \to \sigma(x)$.

One question that immediately emerges is if all instances of self-referential sentences in T are an instance of the Diagonalization Lemma. We will not address this question because, for what we want to argue, we simply need the fact that all sentences that arise from the Diagonalization Lemma are self-referential, and we do not need the converse.

We will assume that self-reference can be defined in T by a predicate $\sigma(x)$ and we will argue that some natural properties about self-reference—for example $\sigma(x) \wedge \mathcal{P}_T(x)$, i.e., being self-referent and

---

[1] $\delta(x) := \exists u.\ (\text{Form}(u) \wedge \exists v.\ v = \ulcorner\forall y(\text{Diag}(z,y) \to \urcorner * s'(u, \ulcorner y\urcorner) * \ulcorner)\urcorner \wedge x = \ulcorner\forall y(\urcorner * \ulcorner\text{Diag}(\dot{v},y)\urcorner * \ulcorner\to\urcorner * s'(u, \ulcorner y\urcorner) * \ulcorner)\urcorner)$ satisfies the desired property.

provable—cannot be totally understood inside T in the sense that they are not decidable using T. More precisely, the next result claims, in particular, that T cannot decide all self-referential sentences that are: provable in T, or provable false in T, or even not provable in T or provably false in T. The previous properties are clearly natural properties about self-referential formulas (keep in mind that a great deal about the discussion about self-reference emerged from Gödel's Incompleteness Theorems, more precisely from the Diagonalization Lemma). The key-aspect of this section is that a proof is given: one might have the intuition that some natural properties related to self-reference, just like provability, might be definable in a theory of Arithmetic but not decidable, but a proof or some kind of argument is needed to sustain such a claim, otherwise is just a supposition.

Inspired by Gödel's Incompleteness Theorems, one might think that the independence phenomenon of the mentioned predicates is coming directly from the fact that the provability predicate is being used, but that is a misleading way of thinking. To illustrate this, one should remember that starting from independent formulas, one can construct decidable formulas using connectives: $\mathcal{G}_T$ is independent, but one has that $\vdash_T \neg(\mathcal{G}_T \wedge \bot)$ and $\vdash_T \mathcal{G}_T \vee \top$. Furthermore, although Gödel's Second Incompleteness Theorem holds in T, the theory T is able to prove the arithmetization of that result [Boo84, p. 61]: $\vdash_T \neg\mathcal{P}_T(\bot) \rightarrow \neg\mathcal{P}_T(\neg\mathcal{P}_T(\bot))$. This confirms that one has no *a priori* reason to think that the use of the provability predicate together with other predicates leads to independence results in every situation. In fact, it is the general formula property that will play a major role in the proof. We now present the independence result.

Theorem 4. *Let $\alpha(x)$ be a general formula. Then the following predicates are not decidable in T:*

1.) $\alpha(x) \wedge \neg\mathcal{P}_T(x)$;

2.) $\neg\alpha(x) \vee \neg\mathcal{P}_T(x)$;

3.) $\alpha(x) \wedge \mathcal{P}_T(\neg x)$;

4.) $\alpha(x) \wedge (\neg \mathcal{P}_T(x) \vee \mathcal{P}_T(\neg x))$;

5.) $\neg \alpha(x) \vee \mathcal{P}_T(x)$;

6.) $\alpha(x) \wedge \mathcal{P}_T(x)$;

7.) $\neg \alpha(x) \vee \neg \mathcal{P}_T(\neg x)$;

8.) $\neg \alpha(x) \vee (\mathcal{P}_T(x) \wedge \neg \mathcal{P}_T(\neg x))$.

Proof. By the proof of Gödel's Second Incompleteness Theorem from [Bar93, p. 828] we know that $\vdash_T \mathcal{G}_T \leftrightarrow \neg \mathcal{P}_T(\ulcorner \bot \urcorner)$; thus $\neg \mathcal{P}_T(\ulcorner \bot \urcorner)$ is independent of T. It suffices to prove that the predicates 1.) up to 4.) are not decidable, because the predicates 5.) up to 8.) are the negations of the previous predicates. Let $\alpha(x)$ be a general formula. We are going to construct independent sentences for the predicates 1.) up to 4.).

1.) Let ψ_1 be the sentence that arises from the application of the Diagonalization Lemma to $\alpha(x) \wedge \neg \mathcal{P}_T(x)$. Then,

$$\vdash_T \psi_1 \leftrightarrow \alpha(\ulcorner \psi_1 \urcorner) \wedge \neg \mathcal{P}_T(\ulcorner \psi_1 \urcorner). \tag{I}$$

Let us see that ψ_1 is independent of T. We have that if $\vdash_T \psi_1$, then

1	$\vdash_T \psi_1$	(Hyp.)
2	$\vdash_T \alpha(\ulcorner \psi_1 \urcorner) \wedge \neg \mathcal{P}_T(\ulcorner \psi_1 \urcorner)$	(I)
3	$\vdash_T \neg \mathcal{P}_T(\ulcorner \psi_1 \urcorner)$	(2)
4	$\vdash_T \mathcal{P}_T(\ulcorner \psi_1 \urcorner)$	(1)
5	$\vdash_T \bot$	(3,4)

which goes against consistency; and if $\vdash_T \neg \psi_1$, then

1	$\vdash_T \neg\psi_1$	(Hyp.)
2	$\vdash_T \neg\alpha(\ulcorner\psi_1\urcorner) \vee \mathcal{P}_T(\ulcorner\psi_1\urcorner)$	(I)
3	$\vdash_T \alpha(\ulcorner\psi_1\urcorner)$	$\alpha(x)$ is a general formula
4	$\vdash_T \mathcal{P}_T(\ulcorner\psi_1\urcorner)$	(2,3)
5	$\vdash_T \mathcal{P}_T(\ulcorner\neg\psi_1\urcorner)$	(1)
6	$\vdash_T \mathcal{P}_T(\ulcorner\psi_1 \wedge \neg\psi_1\urcorner)$	(4,5)
7	$\vdash_T \mathcal{P}_T(\ulcorner\bot\urcorner)$	(6)

which contradicts the proof of Gödel's Second Incompleteness Theorem. Thus, ψ_1 is independent of T, and so $\alpha(x) \wedge \neg\mathcal{P}_T(x)$ is not decidable.

2.) Let ψ_2 be the sentence that arises from the application of the Diagonalization Lemma to $\neg\alpha(x) \vee \neg\mathcal{P}_T(x)$. Then,

$$\vdash_T \psi_2 \leftrightarrow \neg\alpha(\ulcorner\psi_2\urcorner) \vee \neg\mathcal{P}_T(\ulcorner\psi_2\urcorner). \tag{II}$$

Let us prove that ψ_2 is independent of T. We have that if $\vdash_T \psi_2$, then

1	$\vdash_T \psi_2$	(Hyp.)
2	$\vdash_T \neg\alpha(\ulcorner\psi_2\urcorner) \vee \neg\mathcal{P}_T(\ulcorner\psi_2\urcorner)$	(II)
3	$\vdash_T \alpha(\ulcorner\psi_2\urcorner)$	$\alpha(x)$ is a general formula
4	$\vdash_T \neg\mathcal{P}_T(\ulcorner\psi_2\urcorner)$	(2,3)
5	$\vdash_T \mathcal{P}_T(\ulcorner\psi_2\urcorner)$	(1)
6	$\vdash_T \bot$	(4,5)

which goes against consistency, and if $\vdash_T \neg\psi_2$, then

1	$\vdash_T \neg\psi_2$	(Hyp.)
2	$\vdash_T \alpha(\ulcorner\psi_2\urcorner) \wedge \mathcal{P}_T(\ulcorner\psi_2\urcorner)$	(II)
3	$\vdash_T \mathcal{P}_T(\ulcorner\psi_2\urcorner)$	(2)
4	$\vdash_T \mathcal{P}_T(\ulcorner\neg\psi_2\urcorner)$	(1)
5	$\vdash_T \mathcal{P}_T(\ulcorner\psi_2 \wedge \neg\psi_2\urcorner)$	(3,4)
6	$\vdash_T \mathcal{P}_T(\ulcorner\bot\urcorner)$	(5)

which contradicts the proof of Gödel's Second Incompleteness Theorem. Thus, ψ_2 is independent of T, and so $\neg\alpha(x) \vee \neg\mathcal{P}_T(x)$ is not decidable.

3.) Let ψ_3 be the sentence that arises from the application of the Diagonalization Lemma to $\alpha(x) \wedge \mathcal{P}_T(\neg x)$. Then,

$$\vdash_T \psi_3 \leftrightarrow \alpha(\ulcorner\psi_3\urcorner) \wedge \mathcal{P}_T(\ulcorner\neg\psi_3\urcorner). \tag{III}$$

Let us prove that ψ_3 is independent of T. We have that if $\vdash_T \psi_3$, then

1	$\vdash_T \psi_3$	(Hyp.)
2	$\vdash_T \alpha(\ulcorner\psi_3\urcorner) \wedge \mathcal{P}_T(\ulcorner\neg\psi_3\urcorner)$	(III)
3	$\vdash_T \mathcal{P}_T(\ulcorner\neg\psi_3\urcorner)$	(2)
4	$\vdash_T \mathcal{P}_T(\ulcorner\psi_3\urcorner)$	(1)
5	$\vdash_T \mathcal{P}_T(\ulcorner\psi_3 \wedge \neg\psi_3\urcorner)$	(3,4)
6	$\vdash_T \mathcal{P}_T(\ulcorner\bot\urcorner)$	(5)

which goes against the proof of Gödel's Second Incompleteness Theorem; and if $\vdash_T \neg\psi_3$, then

1	$\vdash_T \neg\psi_3$	(Hyp.)
2	$\vdash_T \neg\alpha(\ulcorner\psi_3\urcorner) \vee \neg\mathcal{P}_T(\ulcorner\neg\psi_3\urcorner)$	(III)
3	$\vdash_T \alpha(\ulcorner\psi_3\urcorner)$	$\alpha(x)$ is a general formula
4	$\vdash_T \neg\mathcal{P}_T(\ulcorner\neg\psi_3\urcorner)$	(2,3)
5	$\vdash_T \mathcal{P}_T(\ulcorner\neg\psi_3\urcorner)$	(1)
6	$\vdash_T \bot$	(4,5)

which goes against consistency. Thus, ψ_3 is independent of T, and so $\alpha(x) \wedge \mathcal{P}_T(\neg x)$ is not decidable.

4.) Let ψ_4 be the sentence that arises from the application of the Diagonalization Lemma to $\alpha(x) \wedge (\neg\mathcal{P}_T(x) \vee \mathcal{P}_T(\neg x))$. Then,

$$\vdash_T \psi_4 \leftrightarrow \alpha(\ulcorner\psi_4\urcorner) \wedge (\neg\mathcal{P}_T(\ulcorner\psi_4\urcorner) \vee \mathcal{P}_T(\ulcorner\neg\psi_4\urcorner)). \qquad \text{(IV)}$$

Let us prove that ψ_4 is independent of T. We have that if $\vdash_T \psi_4$, then

1	$\vdash_T \psi_4$	(Hyp.)
2	$\vdash_T \alpha(\ulcorner\psi_4\urcorner) \wedge (\neg\mathcal{P}_T(\ulcorner\psi_4\urcorner) \vee \mathcal{P}_T(\ulcorner\neg\psi_4\urcorner))$	(IV)
3	$\vdash_T \neg\mathcal{P}_T(\ulcorner\psi_4\urcorner) \vee \mathcal{P}_T(\ulcorner\neg\psi_4\urcorner)$	(2)
4	$\vdash_T \mathcal{P}_T(\ulcorner\psi_4\urcorner)$	(1)
5	$\vdash_T \mathcal{P}_T(\ulcorner\neg\psi_4\urcorner)$	(3,4)
6	$\vdash_T \mathcal{P}_T(\ulcorner\psi_4 \wedge \neg\psi_4\urcorner)$	(4,5)
7	$\vdash_T \mathcal{P}_T(\ulcorner\bot\urcorner)$	(6)

which contradicts the proof of Gödel's Second Incompleteness Theorem; and if $\vdash_T \neg\psi_4$, then

1	$\vdash_T \neg\psi_4$	(Hyp.)
2	$\vdash_T \neg\alpha(\ulcorner\psi_4\urcorner) \vee (\mathcal{P}_T(\ulcorner\psi_4\urcorner) \wedge \neg\mathcal{P}_T(\ulcorner\neg\psi_4\urcorner))$	(IV)
3	$\vdash_T \alpha(\ulcorner\psi_4\urcorner)$	$\alpha(x)$ gen. form.
4	$\vdash_T \mathcal{P}_T(\ulcorner\psi_4\urcorner) \wedge \neg\mathcal{P}_T(\ulcorner\neg\psi_4\urcorner)$	(2,3)
5	$\vdash_T \mathcal{P}_T(\ulcorner\psi_4\urcorner)$	(4)
6	$\vdash_T \mathcal{P}_T(\ulcorner\neg\psi_4\urcorner)$	(1)
7	$\vdash_T \mathcal{P}_T(\ulcorner\psi_4 \wedge \neg\psi_4\urcorner)$	(5,6)
8	$\vdash_T \mathcal{P}_T(\ulcorner\bot\urcorner)$	(7)

which goes against the proof of Gödel's Second Incompleteness Theorem. Thus, ψ_4 is independent of T, and so $\alpha(x) \wedge (\neg\mathcal{P}_T(x) \vee \mathcal{P}_T(\neg x))$ is not decidable.

From the previous result we conclude, in particular, that T is not expressible enough to identify self-referential sentences satisfying very natural properties, namely: being provable in T (this corresponds to 6.) in the theorem), being provably false in T (this corresponds to 3.) in the theorem), and not provable in T or provably false in T (this corresponds to 4.) in the theorem). This means that T is not expressible enough to decide some natural properties related to self-reference. As we observed, provability plays a role in the proof but not the major role—the major role is played by the general formulas. To confirm that, one could expect that the predicate in 6.)—the predicate $\alpha(x) \wedge \mathcal{P}_T(x)$—was not decidable because it comes directly from a conjunction with the provability predicate, but one has no *a priori* reason to think that the predicate presented in 2.)—that is equivalent to $\mathcal{P}_T(x) \rightarrow \neg\alpha(x)$—was not decidable.

One immediate consequence of the previous result is the following version of Gödel's First Incompleteness Theorem [Rau06, p. 251].

Corollary 1. *If* Sent(x) *is the predicate that identifies the sentences of T, then the predicate* Sent$(x) \wedge \neg\mathcal{P}_T(x)$ *is not decidable.*

Proof. Follows from 1.) of the previous theorem having in mind that
Sent(x) is a general formula.

◼

3.2 The Diagonalization Lemma and the Strong Diagonalization Lemma: A Separation Result

We start by presenting the Strong Diagonalization Lemma and a
proof of it.

Strong Diagonalization Lemma [Jer73], [Smi13, p. 237]. *For every
one-free-variable-formula $\varphi(x)$ there is a closed-term t such that*

$$\vdash_T t = \ulcorner \varphi(t) \urcorner.$$

Proof. Let $d : \mathbb{N} \to \mathbb{N}$ be the function such that for every one-free-
variable-function-symbol f,

$$d(\#f) := \#\varphi(f(\ulcorner f \urcorner)),$$

and if n is not the code of a one-free-variable-function-symbol, then
$d(n) := 0$. It is clear that d is primitive recursive, so it is captured
by some function-symbol d in T. Hence, given a one-free-variable-
function-symbol f,

$$\vdash_T d(\ulcorner f \urcorner) = \overline{d(\#f)} = \ulcorner \varphi(f(\ulcorner f \urcorner)) \urcorner.$$

In particular, as d is a one-free-variable-function-symbol, we have
that

$$\vdash_T d(\ulcorner d \urcorner) = \ulcorner \varphi(d(\ulcorner d \urcorner)) \urcorner.$$

The conclusion follows by taking t as being the closed-term d($\ulcorner d \urcorner$).

◼

Now we prove that the Diagonalization Lemma does not prove itself: in the sense that instances of $\vdash_T \psi \leftrightarrow \varphi(\ulcorner \psi \urcorner)$ cannot be captured by a single two-free-variable-formula $\alpha(x,y)$ and by applying the Diagonalization Lemma to the one-free-variable-formula $\alpha(\ulcorner \varphi(x) \urcorner, x)$.

Theorem 5. *There is no two-free-variable-formula $\alpha(x,y)$ such that, given a one-free-variable-formula $\varphi(x)$, if ψ is the sentence obtained from the Diagonalization Lemma applied to $\alpha(\ulcorner \varphi(x) \urcorner, x)$, then*

$$\vdash_T \psi \leftrightarrow \varphi(\ulcorner \psi \urcorner).$$

Proof. Suppose, aiming a contradiction, that $\alpha(x,y)$ is in the previous conditions. Take $\varphi(x)$ as being the formula obtained from the Two-Variable Diagonalization Lemma applied to $\neg\alpha(x,y)$. Then,

$$\vdash_T \forall x.\ \varphi(x) \leftrightarrow \neg\alpha(\ulcorner \varphi(x) \urcorner, x),$$

hence

$$\vdash_T \forall x.\ \neg\varphi(x) \leftrightarrow \alpha(\ulcorner \varphi(x) \urcorner, x). \tag{I}$$

Let ψ be the sentence obtained from the Diagonalization Lemma applied to $\alpha(\ulcorner \varphi(x) \urcorner, x)$. Thus,

$$\vdash_T \psi \leftrightarrow \alpha(\ulcorner \varphi(x) \urcorner, \ulcorner \psi \urcorner). \tag{II}$$

As by hypothesis we have

$$\vdash_T \psi \leftrightarrow \varphi(\ulcorner \psi \urcorner), \tag{III}$$

then

1	$\vdash_T \psi \leftrightarrow \alpha(\ulcorner \varphi(x) \urcorner, \ulcorner \psi \urcorner)$	(II)
2	$\vdash_T \psi \leftrightarrow \varphi(\ulcorner \psi \urcorner)$	(III)
3	$\vdash_T \neg\psi \leftrightarrow \neg\varphi(\ulcorner \psi \urcorner)$	(2)
4	$\vdash_T \neg\varphi(\ulcorner \psi \urcorner) \leftrightarrow \alpha(\ulcorner \varphi(x) \urcorner, \ulcorner \psi \urcorner)$	(I)
5	$\vdash_T \neg\psi \leftrightarrow \alpha(\ulcorner \varphi(x) \urcorner, \ulcorner \psi \urcorner)$	(3,4)
6	$\vdash_T \psi \leftrightarrow \neg\psi$	(1,5)
7	$\vdash_T \bot$	(6)

which is a contradiction.

■

The previous result is interesting by its-own because it shows that the Diagonalization Lemma is not powerful enough to prove instances of $\vdash_T \psi \leftrightarrow \varphi(\ulcorner \psi \urcorner)$. As was already observed, one can use the Strong Diagonalization Lemma to derive the Diagonalization Lemma. In an intuitive way, the closed-term from the Strong Diagonalization Lemma is used to construct the sentence ψ as being $\varphi(t)$. One very natural question is if from the sentence ψ from the Diagonalization Lemma one can construct the closed-term t as being $\ulcorner \psi \urcorner$ to prove the Strong Diagonalization Lemma; that is, whether one can use the Diagonalization Lemma to prove the Strong Diagonalization Lemma. The following corollary answers negatively to the question.

Corollary 2. *There is no two-free-variable-formula $\alpha(x,y)$ such that, given a one-free-variable-formula $\varphi(x)$, if ψ is the sentence obtained from the Diagonalization Lemma applied to $\alpha(\ulcorner \varphi(x) \urcorner, x)$, then there is a term t such that ψ is $\varphi(t)$ and*

$$\vdash_T t = \ulcorner \varphi(t) \urcorner.$$

Proof. Suppose, aiming the absurd, that $\alpha(x,y)$ is in the previous conditions. Let $\varphi(x)$ be any one-free-variable-formula, and ψ be the sentence obtained from the Diagonalization Lemma applied to $\alpha(\ulcorner \varphi(x) \urcorner, x)$. Then, there is a term t such that ψ is $\varphi(t)$ and $\vdash_T t = \ulcorner \varphi(t) \urcorner$. We have that

1	$\vdash_T \psi \leftrightarrow \varphi(t)$	(Hyp.)
2	$\vdash_T t = \ulcorner \varphi(t) \urcorner$	(Hyp.)
3	$\vdash_T \psi \leftrightarrow \varphi(\ulcorner \varphi(t) \urcorner)$	(1,2)
4	$\vdash_T \psi \leftrightarrow \varphi(\ulcorner \psi \urcorner)$	(Hyp.) and (3)

Thus, $\alpha(x,y)$ is such that, given a one-free-variable-formula $\varphi(x)$, if ψ is the sentence obtained from the Diagonalization Lemma applied to $\alpha(\ulcorner \varphi(x) \urcorner, x)$, then $\vdash_T \psi \leftrightarrow \varphi(\ulcorner \psi \urcorner)$; which is a contradiction by the previous theorem.

Although from the outside we can see that $\ulcorner \psi \urcorner$ is the closed-term used in the Strong Diagonalization Lemma—that is an arithmetical statement about the equality of two terms—, the Diagonalization Lemma—that is a logical statement about an equivalence of arithmetical formulas—is not expressible enough to do it. Hence, diagonalization of term is substantially different from diagonalization of formulas. This is not totally unexpected fact: since Gödel we know that the Logicism program [Ten17] of reducing Arithmetic to Logic cannot be totally fulfilled. With the previous result we have made explicite one distinction of reasoning with terms (Arithmetic) and reasoning with formulas (Logic).

4 Yablo's Paradox and Self-Reference

Yablo in [Yab93] presented a paradox that, according to him, should not depend on self-reference. Yablo's Paradox arises by considering, for each $n \in \mathbb{N}$, S_n as being the sentence "For all $k > n$, S_k is not true". That is, for each $n \in \mathbb{N}$, we consider

$$S_n \leftrightarrow (\forall k > n.\ \neg S_k). \tag{Y}$$

Let us now see its paradoxical nature[1]. For that, we are firstly going to prove that

$$\forall n \in \mathbb{N}.\ \neg S_n. \tag{I}$$

Let us consider $n \in \mathbb{N}$. Let us suppose, aiming a contradiction, that S_n is the case. So, by (Y), $\forall k > n.\ \neg S_k$ holds. In particular, we have that $\neg S_{n+1}$. Hence, once again by (Y), $\exists k > n + 1.\ S_k$ is the case. Take $m \in \mathbb{N}$ in such conditions. So, S_m holds. But as $m > n + 1 > n$, we conclude, by what was previously seen, that $\neg S_m$ is also the case, which is a contradiction. This proves (I).

Let us continue our derivation of a contradiction. By (I), we have that $\neg S_0$ is the case. So, $\exists k > 0.\ S_k$ holds. Let us consider $\ell > 0$ in such

---

[1]One could argue *a priori* that the paradox arises simply by the fact that the sentences are ill-defined: we need sentences not yet defined to define a particular one like an inverted version of an inductive definition. This problem can be solved by formalising the paradox in Arithmetic and using a general form of the Diagonalization Lemma as exposed in [Pic13]. We will not follow this line of argument because not only we believe that definability is not in the heart of the paradox (since it can be formalised in Arithmetic), but also because we want to step-way from Arithmetic. As we will argue, Yablo's Paradox does not depend on Arithmetic.

© Springer Fachmedien Wiesbaden GmbH, part of Springer Nature 2020
P. G. Santos, *Diagonalization in Formal Mathematics*, BestMasters,
https://doi.org/10.1007/978-3-658-29111-2_4

conditions. Then, S_ℓ is the case. But, by (I), we also have that $\neg S_\ell$ holds, which is a contradiction. This concludes the derivation of a contradiction in the context of Yablo's Paradox.

Several authors have studied Yablo's Paradox in the realm of Arithmetic (especially in PA or similar systems) to discuss its self-referential nature (see [Pri97], [Pic13], [Ket04], and [Coo14]). In what follows, we are going to show that Yablo's Paradox does not depend on Arithmetic by presenting a minimal theory to express it; hence no arithmetical fact is needed to study Yablo's Paradox. From that conclusion, we will argue towards its self-referential nature by interpreting it in Linear Temporal Logic.

4.1 A Minimal Theory to Express Yablo's Paradox

It is clear that the order relation is needed to express Yablo's Paradox. A careful analysis of the derivation of a contradiction that we presented gives the impression that the order relation plays a major role in the derivation and that no arithmetical facts are needed to express Yablo's Paradox. In what follows, we will confirm that impression by expressing Yablo's Paradox in a minimal theory that only deals with a formal version of the relation $<$ and that does not require any Arithmetic.

The theory that we are going to present is very similar to the "scheme (B)" of [Ket04] and to the "theory **Y**" of [HZ17]. The main differences from those two references are mainly the analysis that we are going to make and the conclusion that we are going to draw: we will present a complete formal proof using natural deduction and show that the theory is minimal in a sense yet to explain, and from that we are going to argue that Temporal Logic is the suitable framework to discuss Yablo's Paradox. We agree with the view presented in [HZ17] that Yablo's Paradox does not need the Diagonalization Lemma to be studied. Nevertheless, we go further than what was analysed in the "theory **Y**": in [HZ17] was needed a form of coding (an assignment of constants to formulas) and in our analysis we are not going to use any form of coding, another confirmation that we are not going to need Arithmetic.

We define, as follows, the suitable theory to be considered.

Definition 2. We define the First-Order Theory T as being the theory that has a binary relation symbol $<$ and the following axioms:

AxT1) $\forall x. \forall y. \forall z.\ x < y \wedge y < z \rightarrow x < z$;

AxT2) $\forall x. \exists y.\ x < y$.

The axiom (AxT1) expresses the transitivity of the relation $<$ and the axiom (AxT2) expresses, in an intuitive way, the infinitude of the relation—the actual infinitude would also require (AxT3) $\forall x. \neg x < x$. We decided not to include (AxT3) in the theory T because it is not necessary to express the paradox, but in this context one should conceive this axiom as something very desirable. From (AxT1), it is (AxT2) that fixes the interpretation of the relation (otherwise, in the presence of axiom (AxT3), $<$ could be interpreted as being $>_\mathbb{N}$). Moreover, we have that $\langle \mathbb{N}, <_\mathbb{N} \rangle$ is a model[2] of T. Hence, T is consistent.

Now we move to include a formal version of (Y) in the theory T.

Definition 3. We define the First-Order Theory $\mathcal{Y}$ as being the theory T with a new unary relation symbol S and the axiom:

$$\forall x.\ (S(x) \leftrightarrow (\forall k > x.\ \neg S(k))). \tag{AxY}$$

It is clear that (AxY) formalises (Y) in the theory T. Now we prove the formal version of Yablo's Paradox, i.e., we prove that theory $\mathcal{Y}$ is inconsistent. The proof that we are going to present is a formal version of the derivation presented previously.

Theorem 6. *The theory $\mathcal{Y}$ is inconsistent.*

---

[2]We use the notation $<_\mathbb{N}$ to distinguish the syntactical symbol from the semantical one.

Proof. Let us consider, for each term t of $\mathcal{Y}$, the derivation $\mathcal{D}'_t$ as being

$$
\cfrac{\cfrac{\cfrac{S(t)}{\forall k > t.\ \neg S(k)}\ (\text{AxY})}{\cfrac{\neg S(x_0)}{\neg \forall k > x_0.\ \neg S(k)}\ (\text{AxY})} \qquad x_0 > t}{\exists k > x_0.\ S(k)}
$$

and the derivation $\mathcal{D}''_t$ as being

$$
\cfrac{\cfrac{x_1 > x_0 \wedge S(x_1)}{S(x_1)} \qquad \cfrac{\cfrac{S(t)}{\forall k > t.\ \neg S(k)}\ (\text{AxY}) \qquad \cfrac{x_0 > t \qquad \cfrac{x_1 > x_0 \wedge S(x_1)}{x_1 > x_0}}{x_1 > t}\ (\text{AxT1})}{\neg S(x_1)}}{\bot}
$$

We have the following deduction (let us call it $\mathcal{D}_t$)

$$
\cfrac{\cfrac{(\text{AxT2})}{\cfrac{\forall x. \exists y.\ x < y}{\exists y.\ t < y}\ \forall E} \qquad \cfrac{\cfrac{[S(t)]^1 \qquad [x_0 > t]^2}{\cfrac{\mathcal{D}'_t}{\exists k > x_0.\ S(k)}} \qquad \cfrac{[x_1 > x_0 \wedge S(x_1)]^3 \quad [S(t)]^1 \quad [x_0 > t]^2}{\cfrac{\mathcal{D}''_t}{\bot}\ \exists E_3}}{\bot}\ \exists E_2}{\cfrac{\bot}{\neg S(t)}\ \neg I_1}
$$

Let us consider a fixed term t_0 of $\mathcal{Y}$. Then, we finally have

$$
\cfrac{\cfrac{\cfrac{\mathcal{D}_{t_0}}{\neg S(t_0)}}{\cfrac{\neg S(t_0)}{\neg \forall k > t_0.\ \neg S(k)}\ (\text{AxY})}{\exists k > t_0.\ S(k)} \qquad \cfrac{\cfrac{[x_3 > t_0 \wedge S(x_3)]^4}{S(x_3)} \qquad \cfrac{\mathcal{D}_{x_3}}{\neg S(x_3)}}{\bot}\ \exists E_4}{\bot}
$$

The deduction $\mathcal{D}_t$ corresponds to a formal version of the proof of (I), which confirms that theory $\mathcal{Y}$ is totally suitable to formalise the paradox. Not only theory $\mathcal{Y}$ expresses Yablo's Paradox, but it is also a minimal theory in such conditions (in the sense that if one of the axioms is removed we no longer have inconsistency), as the following result shows.

Theorem 7. *The theories* $\mathcal{Y} - (\text{AxT1})$, $\mathcal{Y} - (\text{AxT2})$, *and* $\mathcal{Y} - (\text{AxY})$ *are consistent.*

Proof. It is clear that $\langle \mathbb{N}, <_{\mathbb{N}} \rangle$ is a model of $\mathcal{Y} - (\text{AxY})$ (here the predicate $S(x)$ can be interpreted in any way). Firstly, let us prove that $\mathcal{Y} - (\text{AxT1})$ is consistent. Let us consider $S = \{0, 1, 2\}$ and $<_S = \{\langle 0, 0 \rangle, \langle 0, 1 \rangle, \langle 1, 0 \rangle, \langle 1, 2 \rangle, \langle 2, 1 \rangle, \langle 2, 2 \rangle\}$. Let us take $S(x)$ as being $x = 1$. Let us see that $\mathfrak{M}_0 = \langle S, <_S, x = 1 \rangle$ is a model of $\mathcal{Y} - (\text{AxT1})$.

It is clear that $\mathfrak{M}_0$ does not satisfy (AxT1) (for instance $0 <_S 1$ and $1 <_S 2$, but it is not the case that $0 <_S 2$). It is straightforward that for each $x \in S$ there is an $y \in S$ such that $x <_S y$. So, (AxT2) is satisfied in $\mathfrak{M}_0$. Now, let us see that (AxY) is satisfied in $\mathfrak{M}_0$. Let us firstly suppose that $x = 1$. We have that $1 <_S 1$ is not the case. So, if $k >_S 1$, then $k \neq 1$. Hence, if $k >_S x$, then $k \neq 1$. Let us now suppose that for each $k \in S$, if $k >_S x$, then $k \neq 1$. So, for all $k \in S$, if $k = 1$, then $k >_S x$ is not the case. This property does not hold for $x = 0$ and $x = 2$ because $1 >_S 0$ and $1 >_S 2$, but it holds for $x = 1$. So, we conclude that $x = 1$. In all, $x = 1$ if, and only if, for each $k \in S$, if $k >_S x$, then $k \neq 1$. Therefore, (AxY) holds in $\mathfrak{M}_0$. In sum, $\mathfrak{M}_0$ is a model of $\mathcal{Y} - (\text{AxT1})$.

Finally, let us prove that $\mathcal{Y} - (\text{AxT2})$ is consistent. Let us see that $\mathfrak{M}_1 = \langle \mathbb{N}, >_{\mathbb{N}}, x = 0 \rangle$ is a model of $\mathcal{Y} - (\text{AxT2})$. It is clear that (AxT2) does not hold in $\mathfrak{M}_1$ and it is also clear that (AxT1) holds in $\mathfrak{M}_1$. Let us see that (AxY) holds in $\mathfrak{M}_1$. Let us firstly suppose that $x = 0$. Then, it is clear that for all $k <_{\mathbb{N}} x$, $k \neq 0$ is the case. Now, let us suppose that for all $k <_{\mathbb{N}} x$, $k \neq 0$ holds. Let us also suppose, aiming a contradiction, that $x \neq 0$. So, $0 <_{\mathbb{N}} x$, which is a contradiction because $0 = 0$. So, $x = 0$. In all, $x = 0$ if, and only if, for all $k <_{\mathbb{N}} x$, $k \neq 0$ is the case. Therefore, (AxY) holds in $\mathfrak{M}_1$. In sum, $\mathfrak{M}_1$ is a model of $\mathcal{Y} - (\text{AxT2})$.

In sum, Yablo's Paradox can be formalised in the (minimal) theory $\mathcal{Y}$ that does not depend on any arithmetical facts and that formalises some properties of $\langle \mathbb{N}, <_\mathbb{N} \rangle$. Hence, Yablo's Paradox depends only in some conditions that the relation $<$ satisfies (transitivity and infinitude), and not on any arithmetical property.

4.2 Linear Time Temporal Logic and Yablo's Paradox

Inspired in the minimal theory $\mathcal{Y}$ and by the fact that Yablo's Paradox does not need Arithmetic to be stated, we are going to establish a connection between Temporal Logic and Yablo's Paradox in order to argue towards its self-referential nature.

Temporal Logic, as the name suggests, is a branch of Logic that aims to study the logical aspects of time and its relation with other logical properties. The most common approach to formalise time is to consider an instant-based model approach to the flow of time. In this context, the flow of time is captured by a set of instants and binary relation of precedence on it, i.e., a pair $\langle T, < \rangle$ (see [GG15] for further details). It is now clear how a relation can be establish between what was concluded using theory $\mathcal{Y}$ and Temporal Logic—the former defines a precedence relation that can be interpreted as being the flow of time by the latter.

As Yablo's Paradox was stated for $\langle \mathbb{N}, <_\mathbb{N} \rangle$, in what follows we are going to consider the Linear Temporal Logic (LTL)[3] that is a temporal

---

[3]A more general Temporal Logic could be considered here, but we decided to study the paradox in LTL because LTL is one of the most well-known Temporal Logics, and as the paradox was stated for $\langle \mathbb{N}, <_\mathbb{N} \rangle$ the formulation in LTL is much clear.

propositional logic interpreted in $\langle \mathbb{N}, <_\mathbb{N} \rangle$ having temporal operators X, G, and F with the following meanings [GG15]:

X :   "In the next moment it will be the case that..."

G :   "It will always be the case that..."

F :   "It will at some time be the case that..."

We consider that F is the dual of G in the sense that, for each formula φ, $F(\varphi)$ can be interpreted as being $\neg G(\neg \varphi)$, and vice-versa. As usual, we drop in every formula unnecessary parenthesis, for example, we write simply $G\varphi$ instead of $G(\varphi)$. The temporal operators X, G, and F are called *nexttime*, *always*, and *sometime*, respectively. The formulas $X\varphi$, $G\varphi$, and $F\varphi$ are usually read "next φ", "allways φ", and "sometime φ" [KM08, pg. 20].

Let us fix, in the remaining section, V a set of propositional variables. The following definition fixes the meaning of the temporal operators.

Definition 4. [KM08, pg. 21] A *temporal* (or *Kripke*) *structure* for V is an infinite sequence $K = \langle \eta_i \rangle_{i \in \mathbb{N}}$, where for each $i \in \mathbb{N}$, $\eta_i : V \to \{0,1\}$. For each K and $i \in \mathbb{N}$ we define $K_i(\varphi)$, where φ is a LTL formula, inductively by:

1.) For each $v \in V$, $K_i(v) = \eta_i(v)$;

2.) $K_i(\bot) = 0$;

3.) $K_i(\varphi \to \psi) = 1$ if, and only if, $K_i(\varphi) = 0$ or $K_i(\psi) = 1$;

4.) $K_i(X\varphi) = K_{i+1}(\varphi)$;

5.) $K_i(G\varphi) = 1$ if, and only if, for every $j \geq i$, $K_j(\varphi) = 1$;

6.) $K_i(F\varphi) = 1$ if, and only if, there is $j \geq i$ such that $K_j(\varphi) = 1$.

For the other logical connectives ($\wedge$, $\vee$, $\leftrightarrow$, and $\neg$), $K_i(\varphi)$ is defined in the usual propositional way.

It is important to observe that in LTL the future includes the present. Now we define validity in LTL.

Definition 5. [KM08, pg. 22] A formula φ is called *valid in the temporal structure K* (or *K satisfies φ*), denoted by $\models_K \varphi$ if, for every $i \in \mathbb{N}$, $K_i(\varphi) = 1$. φ is called a *consequence of* a set $\mathcal{F}$ of formulas, denoted by $\mathcal{F} \models \varphi$, if for every K, $\models_K \varphi$ holds whenever $\models_K \psi$ is the case for all $\psi \in \mathcal{F}$. φ is *valid*, $\models \varphi$, if $\emptyset \models \varphi$.

We define, as follows, the concept of logically equivalent formulas.

Definition 6. [KM08, pg. 23] Two formulas φ and ψ are called *logically equivalent*, $\varphi \cong \psi$, if the formula $\varphi \leftrightarrow \psi$ is valid.

It is important to observe that LTL is an extension of Classical Propositional Logic. From the definition of validity follows directly the following result.

Theorem 8. *The formula* $(\boxplus_0 \cdots \boxplus_n \perp) \to \perp$ *is valid, where, for each* $i \in \{0, \ldots, n\}$, $\boxplus_i \in \{X, G, F\}$.

For what follows, the following valid formulas are relevant (see [KM08]):

(V1) $\neg X\varphi \leftrightarrow X\neg\varphi$;

(V2) $X(\varphi \wedge \psi) \leftrightarrow X\varphi \wedge X\psi$;

(V3) $X(\varphi \leftrightarrow \psi) \leftrightarrow (X\varphi \leftrightarrow X\psi)$;

(V4) $G\varphi \leftrightarrow \varphi \wedge XG\varphi$;

(V5) $\neg G\neg\varphi \leftrightarrow F\varphi$.

The formulas (V1), (V2), and (V3) express some commutativity properties, the formula (V5) expresses the meaning of F, and the formula (V4) expresses the inductive meaning of the future: a formula is always the case if it is valid in the present and if in the next moment it continues always being the case. There is an axiomatisation of

LTL (see [KM08, pg. 33,34]), but we are not going to use it directly; instead, as the previous formulas are valid, we are going to use them to study Yablo's Paradox in LTL.
We now state a very important substitution principle.

Principle of Substitution [KM08, pg. 32]. *For every formulas φ and ψ, if $\varphi \cong \psi$, then $F\varphi \cong F\psi$ and $X\varphi \cong X\psi$.*

Now we move to interpret Yablo's Paradox in LTL. The idea behind what we are going to do is the following: we are going to consider the sentences $S_0, S_1, \ldots$ from Yablo's Paradox as a unique sentence whose truth value is changing in time, this means that we are going to abstract from the particularities of Yablo's Paradox and consider the sentences $S_0, S_1, \ldots$ as a whole. More precisely, we are going to consider a formula S whose truth value changes in time according to condition (Y) from before.

Let us now derive the temporal formulation of Yablo's Paradox. As we are considering a sentence S that changes according to (Y), then we are considering each S_i as being $K_i(S)$ (for every temporal structure K). Hence, we have the following:

$$K_i(S) = 1 \iff \forall j > i. \; K_j(S) = 0,$$

that is,

$$K_i(S) = 1 \iff \forall j \geq i + 1. \; K_j(S) = 0.$$

The previous equivalence in LTL (quantified universally in the temporal structure K) is the formula:

$$S \leftrightarrow XG\neg S. \tag{TY}$$

So, Yablo's Paradox in Temporal Logic is, intuitively, expressed by "This sentence is true if in the next moment it is false in the future" or, if one does not consider that the future includes the present, by "This sentence is true if it is false in the future". One can derive a contradiction in a very similar way as before by thinking (intuitively) about the previous sentences in the natural language. The next result formalises that intuitive approach of time in LTL—if one considers (TY) in LTL one can derive a contradiction, i.e., LTL + (TY) is inconsistent.

Theorem 9. *If the formula* $S \leftrightarrow XG\neg S$ *is valid in LTL, then one can derive* $\perp$ *in LTL.*

Proof.

$$\dfrac{\dfrac{\dfrac{\dfrac{\dfrac{\dfrac{[S]^1}{XG\neg S}\,(TY)}{X(\neg S \wedge XG\neg S)}\,(V4),\text{ Princ. Sub.}}{X\neg S \wedge XXG\neg S}\,(V2)}{X\neg S \wedge XS}\,(TY),\text{Princ. Sub.}}{\neg XS \wedge XS}\,(V1)}{}$$

$$\dfrac{\dfrac{\dfrac{\dfrac{\dfrac{\dfrac{\dfrac{\dfrac{\perp}{\neg S}\,\neg I_1}{\neg XG\neg S}\,(TY)}{X\neg G\neg S}\,(V1)}{XFS}\,(V5),\text{Princ. Sub.}}{XFXG\neg S}\,(TY),\text{Princ. Sub.}}{XFX(\neg S \wedge XG\neg S)}\,(V4),(V3),\text{Princ. Sub.}}{XFX(\neg S \wedge S)}\,(TY),(V3),\text{Princ. Sub.}}{XFX \perp}\,\text{Princ. Sub.}$$

$$\dfrac{XFX \perp}{\perp}\,\text{Theo.8}$$

■

If one translates each step of the previous proof to a formula with quantifiers using Definition 4 one gets the proof of the inconsistency of theory $\mathcal{Y}$ (Theorem 6) or, more intuitively, the original proof of a contradiction presented in Section 4.1. It is important to keep in mind that the concept of time that we are using in LTL is a natural one—it is just a new name for the properties of the relation <. This means that we are not adding an unnatural concept to our formulation of Yablo's Paradox, instead we are simply giving a name to some properties of the considered relation.

Finally, having proved the paradoxical nature of (TY), we argue that the presented formulation of Yablo's Paradox is self-referential. It is clear that this formulation of Yablo's Paradox satisfies the first feature of self-reference presented in the beginning: it has the logical form of feature 1.) from Chapter 2. Intuitively, one might directly be convinced that feature 2.) from the Introduction is satisfied, nevertheless we present a justification. The Liar is commonly accepted to be a self-referential sentence. Our formulation of Yablo's Paradox is more complex than the Liar, because not only the sentence speaks about its very negation (like the Liar), but it also speaks about the future. Hence, if our formulation was not self-referential, being the Liar simpler than it, it would follow that the Liar is not self-referential. This means that feature 2.) is satisfied. In sum, (TY) is a self-referential formulation of Yablo's Paradox, i.e., Yablo's Paradox is self-referential.

5 Smullyan's Theorem, Löb's Theorem, and a General Approach to Paradoxes

In this chapter, we are going to present a result by Smullyan that is behind several important diagonalization phenomena, and we are going to present a general approach to several paradoxes. The main ideas of this chapter are based on a paper by the author: [SK17].

5.1 Smullyan's Theorem

Smullyan in [Smu94] studied several forms of diagonalization, a book where he showed that his Theorem R [Smu94, p. 37] is behind several important diagonalization phenomena [Smu94, p. 25–38] such as: combinatory logic, Gödel's First Incompleteness Theorem, and the Recursion Theorem. We present, as follows, that result.

Theorem R [Smu94, p. 37]. *A sufficient condition for a relation $R(x,y)$ on a set N to have a fixed point is that there be a relation $R'(x,y)$ on N and a function $d : N \to N$ such that:*

(R1) *There is $a \in N$ such that $R'(d(a), a)$;*

(R2) *For each $x, y \in N$, $R'(x, y)$ implies $R(x, d(y))$.*

Proof. Let us consider $a \in N$ as in (R1). We have that $R'(d(a), a)$ so, by (R2), we conclude that $R(d(a), d(a))$. ∎

© Springer Fachmedien Wiesbaden GmbH, part of Springer Nature 2020
P. G. Santos, *Diagonalization in Formal Mathematics*, BestMasters,
https://doi.org/10.1007/978-3-658-29111-2_5

We are going to see that Theorem R not only is behind important results with diagonalization, but also that it is responsible for several paradoxes with diagonalization.

5.2 The Paradoxes and Löb's Theorem

We start by presenting the paradoxes that we are going to study and Löb's Theorem. The Liar Paradox is one of the most ancient paradoxes and certainly one of the most well know. One of its oldest formulations is the Epimenides Paradox:

> "One of their prophets has said, The men of Crete are ever false, evil beasts, lovers of food, hating work."

Titus 1:12

In a simpler form, the Epimenides Paradox can be identified with the sentence: 'Epimenides, a Cretan, says that all Cretans are liars'. It is clear the relation between the former sentence and the sentence: 'This sentence is false'. This last sentence is currently known as the Liar Paradox (or simply the Liar), see [BGR16]. In a logical point of view, this assertion has the following structure: $X \leftrightarrow \neg X$, where X is exactly the sentence that speaks about itself.

Russell's Paradox is a paradox that is deeply related to the Liar. This paradox emerges when the class $Ru = \{x | x \notin x\}$ is considered to be a set. Under that assumption, by definition, we get $Ru \in Ru \leftrightarrow Ru \notin Ru$. It is the later equivalence that is know as Russell's Paradox (see [Mos06],[ID16]). The definition of the class Ru contains a form of diagonalization—the relation $\in (x, y)$ is diagonalized—and Russell's Paradox emerges, by its own, from a diagonalization—the formula $x \in x \leftrightarrow y \notin y$ is diagonalized. Consequently, Russell's Paradox has two diagonalizations.

Another very important paradox, that will serve as basis to comprehend the former paradoxes, is Curry's Paradox [Bea13]. For a sentence φ, Curry's Paradox is the sentence: 'If this sentence is true, then φ'. Let us see, in a symbolic way where X denotes the sentence that speaks about itself, its paradoxal structure:

$$\cfrac{X \leftrightarrow (X \to \varphi)}{\cfrac{\cfrac{X \to (X \to \varphi)}{X \to \varphi}(C) \qquad X \leftrightarrow (X \to \varphi)}{\cfrac{X \to \varphi \qquad\qquad X}{\varphi}}}$$

where (C) denotes the use of the contraction rule. As we can derive any sentence φ, we get a paradox.

Let us now analyse, briefly, Löb's Theorem. Löb in [Löb55] answered the question proposed by Henkin: 'What can we conclude about a similar disposition to Gödel's Theorem without negation?', i.e., 'What can we conclude about a formula B in PA such that $B \leftrightarrow \mathcal{P}_\mathrm{T}(\ulcorner B \urcorner)$?' (confirm [Ver17]). The answer given by Löb is what is nowadays know as Löb's Theorem. Its original formulation (where it is considered the language PA and where $\mathfrak{B}$ denotes the provability predicate in PA) was:

Löb's Theorem (original) [Löb55]. *If $\mathfrak{G}$ is any formula and $\tilde{\mathfrak{B}}(\{\mathfrak{G}\}) \to \mathfrak{G}$ is a theorem, then $\mathfrak{G}$ is a theorem.*

In its most recent formulation, Löb's Theorem is:

Löb's Theorem (adapted from [Bar93]). *Suppose that* T *is a consistent recursive extension of* PA. *Then, if φ is a sentence in* T, *we have*

$$\frac{\mathcal{P}_\mathrm{T}(\ulcorner \varphi \urcorner) \to \varphi}{\varphi}$$

5.3 The Considered Languages and Diagonalization

In what follows, we are going to study Curry's Paradox and its relation with Löb's Theorem and the Liar. For that, we are going to

present a system that generalises Curry's Paradox—the connection
between the system and Curry's Paradox is going to be straightforward because one of the purposes of the system is precisely to
generalise the structure of Curry's Paradox.

As we want to study Löb's Theorem and Russell's Paradox, we
need to allow some form of Gödelization within the system. More
precisely, we are going to consider a first-order theory T such that
for each formula, φ, we can consider its name, $\mathcal{G}[\varphi]$, as a closed-term
within the language[1]. We also want to consider an operator (an unary
meta-predicate), $\Box$, of formulas, φ, within the theory, $\Box\varphi$. We are
going to work under the assumption that T is given and, in some
sense, T can be extended to the predicate $\Box$. The theory T enriched
with[2] $\Box$ will be denoted by $T_\Box$.

We are going to presume that in T is defined a notion of truth or
satisfiability of formulas for formulas where $\Box$ does not occur. In this
chapter, we will denote by $\mathrm{Var}(T)$ the collection of all the variables
of T, by $\mathrm{Term}(T)$ the collection of all terms of T, by $\mathrm{CTerm}(T)$ the
collection of all closed terms in the considered theory, by $\mathrm{Form}(T)$ the
collection of all formulas of T, by $\mathrm{Form}_1(T)$ the collection of all one-
variable formulas of T, and by $\mathrm{Sent}(T)$ the collection of all sentences
of T (the same notations will be considered for $T_\Box$). Having been
introduced these notations, it is important to observe that $\mathcal{G}$ and $\Box$
can be considered as functions, respectively, of types $\mathcal{G} : \mathrm{Form}(T) \rightarrow$
$\mathrm{CTerm}(T)$ and $\Box : \mathrm{Form}(T) \rightarrow \mathrm{Form}(T)$.

For each $\varphi \in \mathrm{Form}(T)$, we will denote by $\mathrm{var}(\varphi)$ the collection of
all free variables of φ. We are going to assume that the following
conditions are satisfied in T:

- In the theory T:

  - *Modus Ponens*

$$\frac{\varphi \rightarrow \psi \quad \varphi}{\psi} \text{ (MP)}$$

---

[1] Here the Gödelization does not need to be the usual one in Arithmetic.
[2] This corresponds to a first-order modal theory.

- Contraction Rule

$$\frac{\varphi \rightarrow (\varphi \rightarrow \psi)}{\varphi \rightarrow \psi} \text{ (C)}$$

- In the meta-theory:

  - Universal Instantiation Rule

    If $\{x_0, \ldots, x_n\} \subseteq \text{var}(\varphi)$, then, for each $t_0, \ldots, t_n \in \text{CTerm}(T)$,

    $$\vdash_T \varphi \implies \vdash_T \varphi \left[\frac{t_0 \ldots t_n}{x_0 \ldots x_n} \right]$$

where $\varphi, \psi \in \text{Form}(T)$. In the Instantiation Rule we are considering the notation of [EFT96]. In what follows, given $\varphi(x_0, \ldots, x_n) \in \text{Form}(T)$ ($x_0, \ldots, x_n$ are, at most, the free variables[3] of φ) and $t_0, \ldots, t_n \in \text{CTerm}(T)$, we will denote $\varphi \left[\frac{t_0 \ldots t_n}{x_0 \ldots x_n} \right]$, as usual, simply by $\varphi(t_0, \ldots, t_n)$.

We now present an adaptation of Theorem R of Smullyan that traces diagonalization to a common reasoning:

Smullyan Theorem (ST) (adapted from [Smu94]). *Given a theory T, a sufficient condition for a formula $R(x, y) \in \text{Form}(T_\square)$ to have a fixed point is that there be a formula $R'(x, y) \in \text{Form}(T_\square)$ and a function-symbol d such that:*

(S1) *There is $t \in \text{CTerm}(T)$ such that $\vdash R'(d(t), t)$;*

(S2) *For each $t_0, t_1 \in \text{CTerm}(T)$, $\vdash R'(t_0, t_1)$ implies $\vdash R(t_0, d(t_1))$.*

Proof. Consider $t \in \text{CTerm}(T)$ in the conditions of (S1). So, $\vdash R'(d(t), t)$ and, by (S2), $\vdash R(d(t), d(t))$.

∎

We trace back every form of diagonalization (used where) to a Smullyan like reasoning. The following result is exactly Smullyan's Theorem where d is the identity-function-symbol:

---

[3]Given a formula φ, we are using the notation $\varphi(x_0, \ldots, x_n)$ with the meaning of $\text{var}(\varphi) \subseteq \{x_0, \ldots, x_n\}$.

Diagonalization Theorem (DT). *Given a theory T, a sufficient condition for a formula $R(x,y) \in \mathrm{Form}(T_\square)$ to have a fixed point is that there be a formula $R'(x,y) \in \mathrm{Form}(T_\square)$ such that:*

(D3) *There is $t \in \mathrm{CTerm}(T)$ such that $\vdash R'(t,t)$;*

(D4) *For each $t_0, t_1 \in \mathrm{CTerm}(T)$, $\vdash R'(t_0, t_1)$ implies $\vdash R(t_0, t_1)$.*

5.4 Curry Systems and Löb Systems

We present, as follows, the general systems that will cover the Liar, Russell's Paradox, Curry's Paradox, and Löb's Theorem.

Definition 7. We say that $\mathcal{C} = \langle T, \square, A(x), B(x), P \rangle$ is a *Curry System*, where $A(x), B(x) \in \mathrm{Form}_1(T) \cup \mathrm{Sent}(T)$ and $P \in \mathrm{Sent}(T)$, if:

C1.) For each $X(x), Y(x) \in \mathrm{Form}_1(T_\square) \cup \mathrm{Sent}(T_\square)$, given $t \in \mathrm{CTerm}(T)$, we have

$$\frac{X(t) \leftrightarrow Y(t)}{\square X(t) \leftrightarrow \square Y(t)} \; (\square \mathrm{I})$$

C2.) $\vdash_{\mathcal{C}} A(\mathcal{G}[A(x)]) \leftrightarrow \square B(\mathcal{G}[A(x)])$;

C3.) For each $t \in \mathrm{CTerm}(T)$, $\vdash_{\mathcal{C}} A(t) \leftrightarrow (\square B(t) \to P)$.

Sometimes we will omit the theory when considering a Curry System: $\mathcal{C} = \langle \square, A(x), B(x), P \rangle$.

Let us analyse the previous definition. The condition C1.) is dispensable for the main result about Curry Systems, but we decided to maintain it in the definition because it is a very natural property to ask to be satisfied, and because it will play a role in the systems that we will define. The operator $\square$ is intended to represent some notion of truth or of provability. The sentence P represents a basis sentence and $A(x)$ and $B(x)$ are formulas that generalise Curry's Paradox layout—in the original Curry's Paradox they are one and the same. We are considering that $\square$ is a priority operator, in the sense that $\square X(x) \leftrightarrow \square Y(x)$ is to be considered as $(\square(X(x))) \leftrightarrow (\square(Y(x)))$. The main result about Curry Systems is the following:

Curry's Theorem. *Let $C = \langle \Box, A(x), B(x), P \rangle$ be a Curry System. We have that $\vdash_C P$. In particular, if $P =\perp$, then $\vdash_C \perp$ (observe that we are not affirming $\perp$ a priori in C).*

Proof.

$$\dfrac{\dfrac{\dfrac{\overset{\text{C2.)}}{A(\mathcal{G}[A(x)]) \leftrightarrow \Box B(\mathcal{G}[A(x)])}}{A(\mathcal{G}[A(x)]) \leftrightarrow (A(\mathcal{G}[A(x)]) \to P)}}{\dfrac{A(\mathcal{G}[A(x)]) \to P}{A(\mathcal{G}[A(x)])} \text{(C)}} \qquad \dfrac{\dfrac{\overset{\text{C3.)}}{A(\mathcal{G}[A(x)]) \leftrightarrow (\Box B(\mathcal{G}[A(x)]) \to P)}}{A(\mathcal{G}[A(x)]) \leftrightarrow (A(\mathcal{G}[A(x)]) \to P)} \text{(Sm.)}}{}}{}$$

$$\dfrac{A(\mathcal{G}[A(x)]) \qquad A(\mathcal{G}[A(x)]) \leftrightarrow (A(\mathcal{G}[A(x)]) \to P)}{\dfrac{A(\mathcal{G}[A(x)]) \qquad A(\mathcal{G}[A(x)]) \to P}{P}}$$

∎

We now turn our attention to the relation between the former proof and the Diagonalization Theorem. For the proof, the diagonalization of the formula $A(y) \leftrightarrow (A(x) \to P)$—that corresponds to the step (Sm.)—is of most importance and we can derive its diagonalization from the Diagonalization Theorem. Let us see how.

Consider $R(x, y)$ as being $A(y) \leftrightarrow (A(x) \to P)$ and $R'(x, y)$ as being $A(x) \leftrightarrow \Box B(y)$. Let us see that the conditions of the Diagonalization Theorem are verified for $R(x, y)$ and $R'(x, y)$:

(D1) The verification of this hypothesis is a part of the previous proof:

$$\dfrac{A(\mathcal{G}[A(x)]) \leftrightarrow \Box B(\mathcal{G}[A(x)])}{R'(\mathcal{G}[A(x)], \mathcal{G}[A(x)])}$$

hence $\vdash_C R'(\mathcal{G}[A(x)], \mathcal{G}[A(x)])$, as wanted, since $\mathcal{G}[A(x)] \in \text{CTerm}(T)$;

(D2) Let $t_0, t_1 \in \text{CTerm}(T)$. Suppose that $\vdash_C R'(t_0, t_1)$. So,

$$\dfrac{\dfrac{R'(t_0, t_1)}{A(t_0) \leftrightarrow \Box B(t_1)} \qquad \dfrac{\overset{\text{C3.)}}{A(t_1) \leftrightarrow (\Box B(t_1) \to P)}}{A(t_1) \leftrightarrow (\Box B(t_1) \to P)}}{\dfrac{A(t_1) \leftrightarrow (A(t_0) \to P)}{R(t_0, t_1)}}$$

Therefore, $\vdash_C R(t_0, t_1)$. By the arbitrariness of t_0 and t_1, we have verified the condition.

The previous theorem is the formal form of Curry's Paradox. The Liar is Curry's Paradox where $P = \bot$. Russell's Paradox follows from Curry's Theorem by considering in the theory of the set theory ZF: $\Box$ as the trivial operator (the trivial meta-predicate), $\mathcal{G}$ as the function that assigns to each formula $\varphi(x_1, \ldots, x_n)$ the closed-term $\{x | \varphi(x, \ldots, x)\}$, $A(x) = x \in \text{Ru}$, $B(x) = x \in x$, and $P = \bot$. Let us analyse what was described.

For Russell's Paradox, the proof of Curry Theorem is simply

$$\frac{\dfrac{A(\mathcal{G}[A(x)]) \leftrightarrow \Box B(\mathcal{G}[A(x)]) \qquad A(\mathcal{G}[A(x)]) \leftrightarrow (\Box B(\mathcal{G}[A(x)]) \to P)}{A(\mathcal{G}[A(x)]) \leftrightarrow (A(\mathcal{G}[A(x)]) \to P)}}{\dfrac{A(\mathcal{G}[A(x)]) \leftrightarrow \neg A(\mathcal{G}[A(x)])}{\bot}}$$

That is, in

$$\frac{\dfrac{\dfrac{\dfrac{A(\mathcal{G}[A(x)]) \leftrightarrow \Box B(\mathcal{G}[A(x)])}{A(\mathcal{G}[A(x)]) \leftrightarrow (A(\mathcal{G}[A(x)]) \to P)} \quad A(\mathcal{G}[A(x)]) \leftrightarrow (\Box B(\mathcal{G}[A(x)]) \to P)}{A(\mathcal{G}[A(x)]) \to P} \text{(Sm.)}}{A(\mathcal{G}[A(x)])} \quad \dfrac{A(\mathcal{G}[A(x)]) \leftrightarrow (A(\mathcal{G}[A(x)]) \to P)}{A(\mathcal{G}[A(x)]) \to P}}{P}$$

it is the highlighted area.

The proof of Curry Theorem for Curry's Paradox is the last section of the proof, i.e., in

$$\frac{\dfrac{\dfrac{\dfrac{A(\mathcal{G}[A(x)]) \leftrightarrow \Box B(\mathcal{G}[A(x)])}{A(\mathcal{G}[A(x)]) \leftrightarrow (A(\mathcal{G}[A(x)]) \to P)} \quad A(\mathcal{G}[A(x)]) \leftrightarrow (\Box B(\mathcal{G}[A(x)]) \to P)}{A(\mathcal{G}[A(x)]) \to P} \text{(Sm.)}}{A(\mathcal{G}[A(x)])} \quad \dfrac{A(\mathcal{G}[A(x)]) \leftrightarrow (A(\mathcal{G}[A(x)]) \to P)}{A(\mathcal{G}[A(x)]) \to P}}{P}$$

corresponds to the highlighted area.

What we have described confirms that Curry systems generalise Curry's Paradox, the Liar, and Russell's Paradox. Furthermore, it confirms that each highlighted area is indispensable for a general system. Aiming the general study of Löb's Theorem and its relation to Curry's Paradox we give the following definition:

Definition 8. We say that $\mathcal{L} = \langle T, \Box, A(x), B(x), P \rangle$ is a *Löb System*, where $A(x), B(x) \in \mathrm{Form}_1(T) \cup \mathrm{Sent}(T)$ and $P \in \mathrm{Sent}(T)$, if:

L1.) (Necessitation) For each $\varphi \in \mathrm{Form}(T_\Box)$,

$$\frac{\varphi}{\Box\varphi}$$

L2.) (Internal Necessitation) For each $\varphi \in \mathrm{Form}(T_\Box)$,

$$\vdash_{\mathcal{L}} \Box\varphi \to \Box\Box\varphi;$$

L3.) (Box Distributivity) For each $\varphi, \psi \in \mathrm{Form}(T_\Box)$,

$$\vdash_{\mathcal{L}} \Box(\varphi \to \psi) \to (\Box\varphi \to \Box\psi);$$

L4.) For each $t \in \mathrm{CTerm}(T)$,

$$\vdash_{\mathcal{L}} A(t) \leftrightarrow (\Box B(t) \to P);$$

L5.) $\vdash_{\mathcal{L}} A(\mathcal{G}[A(x)]) \leftrightarrow \Box B(\mathcal{G}[A(x)]);$

L6.) $\vdash_{\mathcal{L}} A(\mathcal{G}[A(x)]) \leftrightarrow B(\mathcal{G}[A(x)]).$

We will sometimes omit the theory when considering a Löb System: $\mathcal{L} = \langle \Box, A(x), B(x), P \rangle$.

Now, we are going to analyse the relation between Löb Systems and the Peano Arithmetic, PA. Let us consider, for each formula φ, $\mathcal{G}[\varphi] = \ulcorner \varphi \urcorner$ and let us substitute each occurrence of $\Box$, with a certain formula φ, for $\mathcal{P}_T(\ulcorner\varphi\urcorner)$. By the Diagonalization Lemma applied to $\mathcal{P}_{PA}(x)$ (see Chapter 3), we have that there is a sentence φ such that

$$\vdash_{PA} \varphi \leftrightarrow \mathcal{P}_{PA}(\ulcorner\varphi\urcorner).$$

If we take φ in the previous conditions and P as being the sentence $0 = 0$, then $\langle PA, \mathcal{P}_{PA}(\ulcorner\cdot\urcorner), \varphi, \varphi, P\rangle$ obeys to all the conditions that define a Löb System (we cannot say that $\langle PA, \mathcal{P}_{PA}(\ulcorner\cdot\urcorner), \varphi, \varphi, P\rangle$ is a Löb System due to a technical aspect: $\mathcal{P}_{PA}(\ulcorner\cdot\urcorner)$ is not necessarily an operator in the sense that we have been considering $\square$).

We present the relation between Löb Systems and Curry Systems as follows:

Theorem 10. *Every Löb System $\mathcal{L} = \langle\square, A(x), B(x), P\rangle$ is a Curry System $\mathcal{C} = \langle\square, A(x), B(x), P\rangle$.*

Proof. L4.) is, in fact, C3.) and L5.) is C2.). Let us confirm that C1.) is valid. Let $X(x), Y(x) \in \text{Form}_1(T_\square) \cup \text{Sent}(T_\square)$ and $t \in \text{CTerm}(T)$. Then,

$$\cfrac{\cfrac{X(t) \leftrightarrow Y(t)}{\square(X(t) \leftrightarrow Y(t))}\ \text{L1.)}}{\square X(t) \leftrightarrow \square Y(t)}\ \text{L3.)}$$

∎

The following result generalise Theorem 1 of [Lin06] and will allow the proof of a general form of Löb's Theorem.

Fixed Point Theorem. *Let $\mathcal{L} = \langle\square, A(x), B(x), P\rangle$ be a Löb System. We have that there is $\Psi \in \text{Sent}(\mathcal{L})$ such that*

$$\vdash_{\mathcal{L}} \Psi \leftrightarrow (\square\Psi \to P).$$

Proof. Consider $\mathcal{S}$ as being the deduction

$$\cfrac{\cfrac{A(\mathcal{G}[A(x)]) \leftrightarrow \square B(\mathcal{G}[A(x)])}{} \text{L5.)} \qquad \cfrac{A(\mathcal{G}[A(x)]) \leftrightarrow (\square B(\mathcal{G}[A(x)]) \to P)}{} \text{L4.)}}{A(\mathcal{G}[A(x)]) \leftrightarrow (A(\mathcal{G}[A(x)]) \to P)}\ \text{(Sm.)}$$

So,

$$
\cfrac{
  \mathcal{S} \qquad
  \cfrac{
    \cfrac{A(\mathcal{G}[A(x)]) \leftrightarrow \Box B(\mathcal{G}[A(x)])}{}
    \mathrm{L5.)}
    \qquad
    \cfrac{
      \cfrac{
        \cfrac{A(\mathcal{G}[A(x)]) \leftrightarrow B(\mathcal{G}[A(x)])}{\Box(A(\mathcal{G}[A(x)]) \leftrightarrow B(\mathcal{G}[A(x)]))} \mathrm{L6.)}
      }{\Box A(\mathcal{G}[A(x)]) \leftrightarrow \Box B(\mathcal{G}[A(x)])} \mathrm{L1.)\ L3.)}
    }{A(\mathcal{G}[A(x)]) \leftrightarrow \Box A(\mathcal{G}[A(x)])}
  }{A(\mathcal{G}[A(x)]) \leftrightarrow \Box A(\mathcal{G}[A(x)])}
}{A(\mathcal{G}[A(x)]) \leftrightarrow (\Box A(\mathcal{G}[A(x)]) \to P)}
$$

By Theorem 10, we know that every Löb System is a Curry System. So, as each Curry System has the reasoning of the Diagonalization Theorem, then also each Löb System has the reasoning of that theorem. Moreover, the formula $R(x,y)$ that was diagonalised before is also the formula that is considered in this proof, which confirms, once again, the universality and transversality of the reasoning of the Diagonalization Theorem and, consequently, of Smullyan's Theorem. The generalised form of Löb's Theorem is:

Löb's Theorem. *Let $\mathcal{L} = \langle \Box, A(x), B(x), P \rangle$ be a Löb System. Then*

$$\frac{\Box P \to P}{P}$$

Proof. Let us consider $\Psi = A(\mathcal{G}[A(x)])$, $\mathcal{D}$ as being the deduction

$$
\cfrac{
  \cfrac{
    \cfrac{
      \cfrac{\Psi \leftrightarrow (\Box\Psi \to P)}{\Psi \to (\Box\Psi \to P)}
    }{\Box(\Psi \to (\Box\Psi \to P))} \mathrm{L1.)}
  }{\Box\Psi \to \Box(\Box\Psi \to P)} \mathrm{L3.)}
}{}
$$

$\mathcal{D}'$ as being

$$
\cfrac{
  \Box P \to P \qquad
  \cfrac{
    \mathcal{D} \qquad
    \cfrac{\Box(\Box\Psi \to P) \to (\Box\Box\Psi \to \Box P)}{} \mathrm{L3.)}
  }{\Box\Psi \to (\Box\Box\Psi \to \Box P)}
  \qquad
  \cfrac{\Box\Psi \to \Box\Box\Psi}{} \mathrm{L2.)}
}{
  \cfrac{\Box\Psi \to \Box P}{\Box\Psi \to P}
}
$$

By the Fixed Point Theorem, we have that

$$
\cfrac{\cfrac{\mathcal{D}'}{\Box\Psi \to P} \qquad \cfrac{\cfrac{\text{Fixed Point Theo.}}{\Psi \leftrightarrow (\Box\Psi \to P)} \qquad (\Box\Psi \to P) \to \Psi}{\cfrac{\Psi}{\Box\Psi}\ \text{L1.)}}}{P}
$$

As the proof of the Fixed Point Theorem uses the reasoning of the Diagonalization Theorem, even more so the proof of the previous theorem uses the reasoning of the Diagonalization Theorem.

Now we know that Curry's Paradox's layout—Curry System—generalizes Löb's Theorem's layout—Löb System.

5.5 General Systems for Paradoxes

Curry Systems unify the considered paradoxes and Löb's Theorem by means of a common structure. This structure underlies an "infinity" of paradoxes in the sense that we can consider different theories and, for each theory, we can consider different formulas $A(x)$, $B(x)$ and P. Although Curry Systems are very general, one can inquire if different paradoxes can be considered as being a Curry System. There is a simple answer to that: if a certain paradox shares the same structure of the previous paradoxes, being it a propositional paradox or a predicate paradox, it can be fitted into a Curry System. Now we clarify what we mean by sharing the same structure:

Definition 9. Given $\varphi(x_1,\dots,x_n) \in \mathrm{Form}(T_\Box)$ we say that $\mathcal{S}_{\varphi(x_1,\dots,x_n)} = \langle T, \Box, A(x_1), B(x_1), P \rangle$ is a *General System for* $\varphi(x_1,\dots,x_n)$, where $A(x_1)$, $B(x_1) \in \mathrm{Form}_1(T) \cup \mathrm{Sent}(T)$, where $B(x_1)$ is a sub-formula of $\varphi(x_1\dots, x_n)$ and $P \in \mathrm{Sent}(T)$, if:

G1.) For each $X(x), Y(x) \in \mathrm{Form}_1(T_\square) \cup \mathrm{Sent}(T_\square)$, given $t \in \mathrm{CTerm}(T)$, we have

$$\frac{X(t) \leftrightarrow Y(t)}{\square X(t) \leftrightarrow \square Y(t)} \, (\square \mathrm{I})$$

G2.) $\vdash_{\mathcal{S}_{\varphi(x_1,\dots,x_n)}} A(\mathcal{G}[A(x)]) \leftrightarrow \square B(\mathcal{G}[A(x)])$;

G3.) For each $t \in \mathrm{CTerm}(T)$, $\vdash_{\mathcal{S}_{\varphi(x_1,\dots,x_n)}} A(t) \leftrightarrow \varphi(t,\dots,t)$.

The previous definition is a generalisation of the definition of a Curry System—the condition C3.) is generalised for the condition G3.). We say that two types of reasonings (paradoxal or non-paradoxal) share the same structure when they are both a General System for $\varphi(x_1,\dots,x_n)$, for a certain $\varphi(x_1,\dots,x_n) \in \mathrm{Form}(T_\square)$. Curry Systems are the General Systems for $\square B(x) \to P$. The formula $\varphi(x_1,\dots,x_n)$ plays a role similar to the one played by a logical signature—the Liar fits into a Curry System because the negation can be expressed using the implication and $P = \bot$.

For each formula $\varphi(x_1,\dots,x_n)$, the General Systems for $\varphi(x_1,\dots,x_n)$ correspond to new versions of Russell's Paradox. If they are not paradoxal, then they correspond, still, to new versions of Löb's Theorem. If they are paradoxal, they correspond to versions of Curry's Paradox. For each of the paradoxes in [ES08], we can consider a General System that replicates, for different formulas, what was studied here.

5.6 Models of Curry Systems, and Consistency

We recall that we are tacitly assuming that T has some notion of satisfiability for formulas where $\square$ does not occur. The systems that have been considered are similar to a system of modal predicate logic. There are various approaches to the modal predicate logic, in particular to the notion of model in this framework (see, for instance, [Bow79], [Bus98], and [Alo05]). The following definition of model was inspired in the analogous definitions in [Alo05] and [Bus98] (mostly the first one).

Definition 10. $\mathfrak{M} = \langle W, R, \langle \mathcal{A}_w \rangle_{w \in W} \rangle$ is a *structure system* for T if W is a nonempty set, if R is a reflexive and transitive relation in W, if, for each $w \in W$, $\mathcal{A}_w = \langle A_w, F_w \rangle$ is a T-structure ([Bar93, p. 17]), and if, for each $u, v \in W$ such that $R(u, v)$, the following properties are verified[4]:

St.1.) $A_u \subseteq A_v$;

St.2.) If $t \in \mathrm{Term}(T)$ and $s : \mathrm{Var}(T) \to A_u$ is an attribution (assignment)[5], then[5] $t^{\mathcal{A}_u}[s] = t^{\mathcal{A}_v}[s]$;

St.3.) If S is a relation-symbol of T, then[6] $S^{\mathcal{A}_u} = S^{\mathcal{A}_v}$.

Given $w \in W$ and an attribution $s : \mathrm{Var}(T) \to A_w$, we define inductively the relation $w \Vdash \varphi[s]$ for each formula φ of $T_\square$ by[7]:

1.) If $\square$ does not occur in φ, then $w \Vdash \varphi[s] \iff \mathcal{A}_w \models \varphi[s]$;

2.) $w \Vdash (\square\varphi)[s] \iff (\mathbb{V}w' \in W.R(w,w') \implies w' \Vdash \varphi[s])$;

3.) $w \Vdash (\varphi \wedge \psi)[s] \iff (w \Vdash \varphi[s]) \& (w \Vdash \varphi[s])$;

4.) $w \Vdash (\varphi \vee \psi)[s] \iff (w \Vdash \varphi[s]) \bigvee (w \Vdash \varphi[s])$;

5.) $w \Vdash (\neg\varphi)[s] \iff \sim (w \Vdash \varphi[s])$;

6.) $w \Vdash (\varphi \to \psi)[s] \iff ((w \Vdash \varphi[s]) \implies (w \Vdash \varphi[s]))$;

7.) $w \Vdash (\varphi \leftrightarrow \psi)[s] \iff ((w \Vdash \varphi[s]) \iff (w \Vdash \varphi[s]))$;

8.) $\sim (w \Vdash \perp [s])$;

9.) $w \Vdash (\forall x.\varphi)[s] \iff \mathbb{V}a \in A_w.w \Vdash \varphi\left[s\binom{x}{a}\right]$;

---

[4]In what follows, in this chapter, we will use the following symbols to denote the logical symbols for the natural language: $\implies$ (implication), $\iff$ (equivalence), & (conjunction), $\bigvee$ (disjunction), $\mathbb{V}$ (universal quantification), and $\exists$ (existential quantification).

[5]See [Bar93, p. 20].

[6]See [Bar93, p. 17].

[7]See [Bar93, p. 21].

10.) $w \Vdash (\exists x.\varphi)[s] \iff \exists a \in A_w.w \Vdash \varphi\left[s\binom{x}{a}\right].$

Given $w \in W$, we write $w \Vdash \varphi$ if for every attribution $s : \mathrm{Var}(T) \to A_w$, $w \Vdash \varphi[s]$. Furthermore, we say that φ is *valid* in $\mathfrak{M}$, and write $\mathfrak{M} \models \varphi$ if, for each $w \in W$, $w \Vdash \varphi$. We say that $\mathfrak{M}$ is a model for a Curry System $\mathcal{C}$ and write $\mathfrak{M} \models \mathcal{C}$ (respectively, a model for a Löb System $\mathcal{L}$, $\mathfrak{M} \models \mathcal{L}$) if for each theorem φ of the system, $\mathfrak{M} \models \varphi$. We say that a Curry System $\mathcal{C}$ (respectively, a Löb System $\mathcal{L}$) is consistent if there exists a model for $\mathcal{C}$ (respectively, for $\mathcal{L}$).

Theorem 11. *Let* $\mathfrak{M} = \langle W, R, \langle A_w \rangle_{w \in W} \rangle$ *be a structure system for* T. *We have that the Converse Barcan formula is valid*[8]:

$$(\Box \forall x.\varphi) \to (\forall x.\Box\varphi).$$

Proof. Let $w \in W$. Consider an attribution $s : \mathrm{Var}(T) \to A_w$ and suppose that $w \Vdash (\Box \forall x.\varphi)[s]$. So, by 2.), we have that $\forall w' \in W.(R(w, w') \implies w' \Vdash (\forall x.\varphi)[s])$ and by 9.),

$$\forall w' \in W.\left(R(w, w') \implies \left(\forall a \in A_{w'}.w' \Vdash \varphi\left[s\binom{x}{a}\right]\right)\right). \tag{I}$$

Let us prove that $w \Vdash (\forall x.\Box\varphi)[s]$, which, by 9.), is equivalent to $\forall a \in A_w.w \Vdash \Box\varphi\left[s\binom{x}{a}\right]$, that, by 2.), is

$$\forall a \in A_w.\left(\forall w' \in W.\left(R(w, w') \implies w' \Vdash \varphi\left[s\binom{x}{a}\right]\right)\right).$$

Let us take $a \in A_w$ and $w' \in W$ such that $R(w, w')$. As $R(w, w')$, we have that $A_w \subseteq A_{w'}$, and so $a \in A_{w'}$. By (I), we conclude that $w' \Vdash \varphi\left[s\binom{x}{a}\right]$, as wanted. So, $w \Vdash ((\Box \forall x.\varphi) \to (\forall x.\Box\varphi))[s]$. By the arbitrary choice

---

[8]This result is a confirmation that our notion of structure is sound. Intuitively, the result claims something that holds in the case where $\Box$ is interpreted as being the provability predicate in PA: if $\vdash_{PA} \forall x.\varphi$, then for each natural number n, $\vdash_{PA} \varphi\left[\frac{n}{x}\right]$.

of s, we have that $w \Vdash ((\Box \forall x. \varphi) \rightarrow (\forall x. \Box \varphi))$. Furthermore, by the arbitrary choice of $w \in W$ we have that $\mathfrak{M} \models ((\Box \forall x. \varphi) \rightarrow (\forall x. \Box \varphi))$.

∎

Now we prove a very important result: a consistency result for certain Curry Systems and Löb Systems.

Consistency Theorem. *If T is consistent, if $A(x)$, $B(x) \in \mathrm{Form}_1(T) \cup \mathrm{Sent}(T)$, $P \in \mathrm{Sent}(T)$, and if $A(x)$, $B(x)$ and P are provable (hence true) in T, then $\mathcal{L} = \langle T, \Box, A(x), B(x), P \rangle$ (respectively, $\mathcal{C} = \langle T, \Box, A(x), B(x), P \rangle$) is consistent.*

Proof. Let us suppose that T is consistent, that $A(x)$, $B(x) \in \mathrm{Form}_1(T) \cup \mathrm{Sent}(T)$, $P \in \mathrm{Sent}(T)$, and that $A(x)$, $B(x)$ and P are provable in T. Let $\mathfrak{N} = \langle N, F \rangle$ be a model of T. We have that $\mathfrak{N} \models A(x)$, $\mathfrak{N} \models B(x)$, and $\mathfrak{N} \models P$. Let us consider $W = \mathrm{Term}(T)$, R the identity relation in W, and, for each $w \in W$, $A_w = \mathfrak{N}$. Furthermore, consider $\mathfrak{M} = \langle W, R, \langle A_w \rangle_{w \in W} \rangle$. We have that $\mathfrak{M}$ is a structure system for T.

Take $\mathcal{L} = \langle T, \Box, A(x), B(x), P \rangle$. Let us prove that $\mathfrak{M} \models \mathcal{L}$.

L1.) Let us suppose that $\mathfrak{M} \models \varphi$. Consider $w \in W$ and $s : \mathrm{Var}(T) \rightarrow A_w$ an attribution. By hypothesis, $w \Vdash \varphi[s]$. Let $w' \in W$ be such that $R(w, w')$. Hence, $w = w'$, and so $w' \Vdash \varphi[s]$. We can, therefore, conclude that $\mathbb{V}w' \in W.(R(w, w') \implies w' \Vdash \varphi[s])$, which means that $w \Vdash (\Box \varphi)[s]$. By the arbitrariness of s we have that $w \Vdash \Box \varphi$, and by the arbitrariness of w we have that $\mathfrak{M} \models \Box \varphi$.

L2.) Let $w \in W$ and $s : \mathrm{Var}(T) \rightarrow A_w$ be an attribution. Suppose that $w \Vdash (\Box \varphi)[s]$. So,

$$\mathbb{V}w' \in W.(R(w, w') \implies w' \Vdash \varphi[s]).$$

Let us prove that $w \Vdash (\Box \Box \varphi)[s]$, which is equivalent to

$$\mathbb{V}w' \in W.(R(w, w') \implies w' \Vdash \Box \varphi[s]),$$

that is

$$\mathbb{V}w' \in W.((R(w, w') \implies (\mathbb{V}w'' \in W.(R(w', w'') \implies w'' \Vdash \varphi[s])))).$$

For that purpose, consider $w', w'' \in W$ such that $R(w, w')$ and $R(w', w'')$. As R is transitive, we have that $R(w, w'')$. So $w = w''$, from which we conclude that $w'' \Vdash \varphi[s]$. Therefore, $w \Vdash (\Box\Box\varphi)[s]$. So, $w \Vdash (\Box\varphi \to \Box\Box\varphi)[s]$. By the arbitrariness of s and w, we have that $\mathfrak{M} \models (\Box\varphi \to \Box\Box\varphi)$.

L3.) Take $w \in W$ and $s : \mathrm{Var}(T) \to A_w$ an attribution. Let us suppose that
$w \Vdash (\Box(\varphi \to \psi))[s]$. So,

$$\forall w' \in W.(R(w, w') \implies w' \Vdash (\varphi \to \psi)[s]),$$

and consequently

$$\forall w' \in W.(R(w, w') \implies (w' \Vdash \varphi[s] \implies w' \Vdash \psi[s])). \tag{I}$$

Let us prove that $w \Vdash (\Box\varphi \to \Box\psi)[s]$. Suppose that $w \Vdash (\Box\varphi)[s]$, that is

$$\forall w' \in W.(R(w, w') \implies w' \Vdash \varphi[s]). \tag{II}$$

Let $w' \in W$ be such that $R(w, w')$. So, as by (I) we have that $w' \Vdash \varphi[s] \implies w' \Vdash \psi[s]$ and by (II) we have that $w' \Vdash \varphi[s]$, we can conclude that $w' \Vdash \psi[s]$. By the arbitrariness of w' we conclude that $w \Vdash (\Box\psi)[s]$. All this means that $w \Vdash (\Box\varphi \to \Box\psi)[s]$ and, by the arbitrariness of w and s, $\mathfrak{M} \models (\Box(\varphi \to \psi) \to (\Box\varphi \to \Box\psi))$.

L4.) Consider $a \in \mathrm{CTerm}(T)$. As we observed before, we have that $\mathfrak{R} \models A(x)$, $\mathfrak{R} \models B(x)$, and $\mathfrak{R} \models P$. So, $\mathfrak{R} \models A(a)$, $\mathfrak{R} \models B(a)$. As $A(x), B(x)$ and P are formulas of $\mathcal{L}$, this means that $\mathfrak{M} \models A(a)$, $\mathfrak{M} \models B(a)$, and $\mathfrak{M} \models P$. By L1.), $\mathfrak{M} \models \Box B(a)$. Let $w \in W$ and $s : \mathrm{Var}(T) \to A_w$ be an attribution. In particular, we have that $w \Vdash A(a)[s]$, $w \Vdash \Box B(a)[s]$, and $w \Vdash P[s]$. So, $w \Vdash (\Box B(a) \to P)[s]$. Therefore, $w \Vdash (A(a) \leftrightarrow (\Box B(a) \to P))[s]$. By the arbitrariness of w and s, $\mathfrak{M} \models (A(a) \leftrightarrow (\Box B(a) \to P))$.

L5.), L6.) By what was seen in L4.), given $a \in \mathrm{CTerm}(T)$, $\mathfrak{M} \models A(a)$, $\mathfrak{M} \models B(a)$, and $\mathfrak{M} \models \Box B(a)$. In particular, $\mathfrak{M} \models A(\mathcal{G}[A(x)])$, $\mathfrak{M} \models B(\mathcal{G}[A(x)])$, and $\mathfrak{M} \models \Box B(\mathcal{G}[A(x)])$. Let $w \in W$ and $s : \mathrm{Var}(T) \to A_w$ be an attribution.

We have that $w \Vdash A(\mathcal{G}[A(x)])[s]$, $w \Vdash B(\mathcal{G}[A(x)])[s]$, and $w \Vdash (\Box B(\mathcal{G}[A(x)]))[s]$. So, $w \Vdash (A(\mathcal{G}[A(x)]) \leftrightarrow B(\mathcal{G}[A(x)]))[s]$ and $w \Vdash (A(\mathcal{G}[A(x)]) \leftrightarrow \Box B(\mathcal{G}[A(x)]))[s]$. By the arbitrariness of w and s, we conclude that $\mathfrak{M} \models (A(\mathcal{G}[A(x)]) \leftrightarrow \Box B(\mathcal{G}[A(x)]))$, and $\mathfrak{M} \models (A(\mathcal{G}[A(x)]) \leftrightarrow B(\mathcal{G}[A(x)]))$.

The proof of the consistency of $\mathcal{C} = \langle T, \Box, A(x), B(x), P \rangle$ is similar.

∎

The previous result is very important because it shows that there are some consistent situations, just like Löb's Theorem; which avoids the trivialities of inconsistent systems.

In sum, Curry's Paradox layout generalises the Liar, Russell's Paradox, general forms of diagonalization like the Liar, and Löb's Theorem reasoning. Where diagonalization is used, one can apply the reasoning of Smullyan's Theorem. If the sentence P in Curry's Paradox is unprovable, then we get something like the Liar, if P is provable, we are in a Löb's Theorem's like layout. There are consistent situations where a reasoning like Curry's Paradox is valid (Consistency Theorem).

6 General Theory of Diagonalization

In this chapter we will present a general theory of diagonalization. We will move towards the goal of the next chapter: the study of Mathematical examples.

Just like the confirmation that ZF can capture the majority of Mathematics comes from examples, also the theory that we will present is confirmed to be behind the majority of important diagonalization phenomenon by the study of relevant Mathematical examples. We start by presenting the language of the general theory.

Definition 11. Let $\mathcal{D}_0$ be the First-Order Language having a binary function-symbol $\circ$, and a unary relation-symbol W. We use the convention that st denotes $s \circ t$ and we define:

$$t : W := W(t);$$
$$\exists x : W. \, A := \exists x. \, (x : W \wedge A);$$
$$\forall x : W. \, A := \forall x. \, (x : W \rightarrow A);$$
$$t : W \rightarrow W := \forall x : W. \, (tx : W);$$
$$t : (W \rightarrow W) \rightarrow (W \rightarrow W) := \forall x : W \rightarrow W. \, (tx : W \rightarrow W).$$

It is important to keep in mind that the intuitive interpretation of $\circ$ is that of function application. Now we define the general theory.

Definition 12. We say that $d = \langle T, \varphi, \varphi' \rangle$ is a *diagonal theory* if T is a First-Order Theory whose language includes $\mathcal{D}_0$, if φ is a two-free-variable-formula, φ' is a formula that has at most four free-variables, and if it has the axiom:

$$\exists d : (W \rightarrow W) \rightarrow (W \rightarrow W). \, (\text{AxD1}(d) \wedge \text{AxD2}(d)), \qquad (\text{AxD})$$

© Springer Fachmedien Wiesbaden GmbH, part of Springer Nature 2020
P. G. Santos, *Diagonalization in Formal Mathematics*, BestMasters,
https://doi.org/10.1007/978-3-658-29111-2_6

where

$\mathsf{AXD1}(d) := \exists F : W \to W. \exists x. \forall y. \varphi'(dF, F, x, y);$

$\mathsf{AXD2}(d) := \forall G : W \to W. \forall H : W \to W. \forall x. (\varphi'(G, H, x, x) \to \varphi(Gx, dHx)).$

The naïve idea behind the previous definition is that it is being used a form of double diagonalization: firstly, there is a meta diagonalization of the "functions" (the intuitive interpretation of elements $f : W \to W$ is that of a function)—using $d : (W \to W) \to (W \to W)$ and axiom AxD1—; and finally there is a diagonalization at the level of terms—captured by the axiom AxD2 that allows the carry of the diagonalization from axiom AxD1. We present, as follows, the confirmation that a diagonal theory $d = \langle T, \varphi, \varphi' \rangle$ proves the diagonalization of φ as was previously described in an intuitive way.

General Diagonalization Theorem (GDT). *Let* $d = \langle T, \varphi, \varphi' \rangle$ *be a diagonal theory. Then,*
$\vdash_T \exists z. \varphi(z, z).$

Proof.

1	$\exists d : (W \to W) \to (W \to W). (\mathsf{AxD1}(d) \land \mathsf{AxD2}(d))$	(AxD)
2	$\mathsf{AxD1}(d_0) \land \mathsf{AxD2}(d_0)$	(Hyp. d_0)
3	$\exists F : W \to W. \exists x. \forall y. \varphi'(d_0 F, F, x, y)$	AxD1(d_0)
4	$\forall G : W \to W. \forall H : W \to W. \forall x. (\varphi'(G, H, x, x) \to \varphi(Gx, d_0 Hx))$	AxD2(d_0)
5	$\exists x. \forall y. \varphi'(d_0 F_0, F_0, x, y)$	(Hyp. F_0)
6	$\forall y. \varphi'(d_0 F_0, F_0, x_0, y)$	(Hyp. x_0)
7	$\varphi'(d_0 F_0, F_0, x_0, x_0)$	$\forall$E (6)
8	$\varphi'(d_0 F_0, F_0, x_0, x_0) \to \varphi(d_0 F_0 x_0, d_0 F_0 x_0)$	$\forall$ E (4)
9	$\varphi(d_0 F_0 x_0, d_0 F_0 x_0)$	$\to$E (7,8)
10	$\exists z. \varphi(z, z)$	$\exists$I (9)
11	$\exists z. \varphi(z, z)$	$\exists$E (5,7–10)
12	$\exists z. \varphi(z, z)$	$\exists$E (3,5–11)
13	$\exists z. \varphi(z, z)$	$\exists$E (1,2–12)

∎

6.1 Towards the Study of Diagonalization in Mathematics

Now we move towards the application of GDT to "everyday Mathematics". The following result is the model version of the GDT and follows immediately from it.

Theorem 12. *A sufficient condition for a binary relation R on a set S to have a fix point is that there be a relation R' such that R and R' are relations on a structure $\mathfrak{M}$ that is a model of a diagonal theory d, where R is interpreting φ, and where R' is interpreting φ'.*

Now we present a more explicit version of the previous result.

Structural General Diagonalization Theorem (SGDT). *A sufficient condition for a binary relation R on a set S to have a fix point is that there be a relation $R' \subseteq S^S \times S^S \times S \times S$ and a function $d : (S \to S) \to (S \to S)$ such that:*

(SGDT1) *There is a function $F : S \to S$ and $x \in S$ such that for all $y \in S$, $R'(d(F), F, x, y)$;*

(SGDT2) *For all function $G, H : S \to S$ and for all $x \in S$, if $R'(G, H, x, x)$, then $R(G(x), (d(H))(x))$.*

Proof. Let $F : S \to S$ and $x \in S$ be in the conditions of (SGD1). Then, for all $y \in S$, $R'(d(F), F, x, y)$ holds; in particular, $R'(d(F), F, x, x)$ holds. By (SGD2) follows that $R((d(F))(x), (d(F))(x))$, as wanted.

∎

The previous proof corresponds to a model-theoretic version of the formal proof of the GDT. Now we present yet another version of the GDT: this time it is a Category Theory version.

Category Theory General Diagonalization Theorem (CTGDT). *In a category C, given an object S, a sufficient condition for a binary relation $R \subseteq \mathrm{Hom}(1, S) \times \mathrm{Hom}(1, S)$ to have a fix point is that there be objects A, B, a relation $R' \subseteq \mathrm{Hom}(A, S) \times \mathrm{Hom}(A, B) \times \mathrm{Hom}(1, A) \times \mathrm{Hom}(1, A)$, and a function $d : \mathrm{Hom}(A, B) \to \mathrm{Hom}(A, S)$ such that:*

(CTGD1) *There are morphisms* $f : A \to B$ *and* $x : 1 \to A$ *such that for all* $y : 1 \to A$,

$R'(d(f), f, x, y)$;

(CTGD2) *For all morphisms* $g : A \to S$, $h : A \to B$ *and for all* $x : 1 \to A$, *if* $R'(g, h, x, x)$, *then* $R(xg, xd(h))$.

Proof. Let $f : A \to B$ and $x : 1 \to A$ be in the conditions of (CTGD1). Then, for all $y : 1 \to A$, $R'(d(f), f, x, y)$. In particular, $R'(d(f), f, x, x)$ is the case. But by (CTGD2) follows that $R(xd(f), xd(f))$, as wanted. ∎

To end this chapter we prove that Theorem R is a particular case of the SGDT: we recall that Theorem R was heavily used in Chapter 5.

Theorem R (TR). *A sufficient condition for a relation* $R(x, y)$ *on a class* N *to have a fixed point is that there be a relation* $R'(x, y)$ *on* N *and a function* $d : N \to N$ *such that:*

(R1) *There is* $a \in N$ *such that* $R'(d(a), a)$;

(R2) *For each* $x, y \in N$, $R'(x, y)$ *implies* $R(x, d(y))$.

Proof. Let $d_1 : (N \to N) \to (N \to N)$ be the function given by

$$d_1(F) := d \circ F.$$

Let $R_1'(G, H, x, y)$ be the relation $R'(G(x), H(x))$. Let us see that the conditions of the SGDT are satisfied for R, R_1', and d_1:

(SGDT1) By hypothesis, there is $a \in N$ such that $R'(d(a), a)$. So, there is $a \in N$ such that $R'((d \circ id_N)(a), id_N(a))$ (where id_N denotes the identity function on N), i.e., there is $a \in N$ such that $R_1'(d_1(id_N), id_N, a, y)$. Hence, there is a function $F : S \to S$ and $x \in S$ such that for all $y \in S$, $R_1'(d_1(F), F, x, y)$.

(SGDT2) Let $G, H : S \to S$ be functions and $x \in S$. Let us suppose that $R_1'(G, H, x, x)$ holds. Then, by hypothesis, we have that $R'(G(x), H(x))$, and thus $R(G(x), d(H(x)))$. So, we have that $R(G(x), (d \circ H)(x))$, i.e., $R(G(x), (d_1(H))(x))$. In sum, for all functions $G, H : S \to S$ and for all $x \in S$, if $R_1'(G, H, x, x)$, then $R(G(x), (d_1(H))(x))$.

By the SGDT we have that R has a fixed point.

∎

7 Mathematical Examples

In this chapter we will study several examples of diagonalization from Mathematics and we will show that all of them are a particular case of the reasoning of the GDT.

7.1 Lawvere's Diagonal Argument

In this section we will show that the main result of the paper [Law69]—1.1.Theorem—is a particular case of the CTGDT. We will also see that the two main theorems of [YM03] are a consequence of the main result from [Law69]. We will follow the notions introduced in [Law69].

We start by recalling that a cartesian closed category is a category **C** equipped with the following three kinds of right-adjoints: a right adjoint 1 to the unique

$$\mathbf{C} \longrightarrow 1,$$

a right-adjoint × to the diagonal functor

$$\mathbf{C} \longrightarrow \mathbf{C} \times \mathbf{C},$$

and for each object A in **C**, a right-adjoint $(\ )^A$ to the functor

$$\mathbf{C} \xrightarrow{\ A \times (\ )\ } \mathbf{C}.$$

© Springer Fachmedien Wiesbaden GmbH, part of Springer Nature 2020
P. G. Santos, *Diagonalization in Formal Mathematics*, BestMasters,
https://doi.org/10.1007/978-3-658-29111-2_7

The adjoint transformations for these adjoint situations, assumed given, will be denoted by λ_A, ϵ_A in our case of exponentiation by A. Thus, for each X we have

$$AX \xrightarrow{X\lambda_A} (A \times X)^A$$

and for each Y

$$A \times Y^A \xrightarrow{Y\epsilon_A} Y.$$

Given $f : A \times X \to Y$, the composite morphism

$$X \xrightarrow{X\lambda_A} (A \times X)^A \xrightarrow{f^A} Y^A$$

will be called the λ-*transform* of the morphism f. A morphism h : $X \to Y^A$ is the λ-transform of f if, and only if, the following diagram commutes

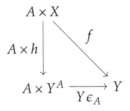

In particular, f is uniquely recovered from its λ-transform. For the particular case of $X = 1$, every $f : A \to Y$ gives rise to a unique $\ulcorner f \urcorner : 1 \to Y^A$ and every $1 \to Y^A$ is of that very form for a unique f. Since for every $a : 1 \to A$ one has, dropping the indices A, Y on ϵ,

$$\langle a, \ulcorner f \urcorner \rangle \epsilon = af.$$

We say that

$$X \xrightarrow{\quad g \quad} Y^A$$

is *weakly point-surjective* if for every $f : A \to Y$ there is x such that for all $a : 1 \to A$,

$$\langle a, xg \rangle \epsilon = af.$$

Furthermore, we say that an object Y has the *fixed point property* if for every endomorphism $t : Y \to Y$ there is $y : 1 \to Y$ such that $y.t = y$.

Lawvere Theorem [Law69]. *In any cartesian closed category, if there is an object A and a weakly point-surjective morphism*

$$A \xrightarrow{\quad g \quad} Y^A$$

then Y has the fixed point property.

Proof. Let $t : Y \to Y$ be a fixed morphism. Given a morphism $f : A \to Y^A$, let $\bar{f}$ be the morphism whose λ-transform is f. We have that for every $f : A \to Y$ there is $x : 1 \to A$ such that for all $a : 1 \to A$

$$\langle a, x \rangle \bar{g} = af.$$

Let us consider $R' \subseteq \mathrm{Hom}(A, Y) \times \mathrm{Hom}(A, Y^A) \times \mathrm{Hom}(1, A) \times \mathrm{Hom}(1, A)$ be the relation such that $R'(f, h, x, y)$ if, and only if,

$$\langle x, y \rangle \bar{h} = xft.$$

Let us also consider $R \subseteq \mathrm{Hom}(1, Y) \times \mathrm{Hom}(1, Y)$ be the relation such that $R(x, y)$ if, and only if, $xt = y$; and the function $d : \mathrm{Hom}(A, Y^A) \to \mathrm{Hom}(A, Y)$ given by

$$d(g) = (A\delta)\bar{g}.$$

Let us see that the conditions of the CTGDT are satisfied:

(CTGD1) Let us consider the composition

$$A \xrightarrow{\ A\delta\ } A \times A \xrightarrow{\ \overline{g}\ } Y \xrightarrow{\ t\ } Y.$$

There is an $x : 1 \to A$ such that for all $a : 1 \to A$

$$\langle a, x \rangle \overline{g} = a(A\delta)\overline{g}t,$$

that is,

$$\langle a, x \rangle \overline{g} = ad(g)t;$$

which is equivalent, by definition, to $R'(d(g), g, x, a)$. Hence, there are $f : A \to Y^A$ and $x : 1 \to A$ such that for all $y : 1 \to A$, $R'(d(f), f, x, y)$ holds.

(CTGD2) Let us consider morphisms $f : A \to Y$, $h : A \to Y^A$, and $x : 1 \to A$. Furthermore, let us suppose that $R'(f, h, x, x)$ holds. Then,

$$\langle x, x \rangle \overline{h} = xft,$$

and so

$$x(A\delta)\overline{h} = xft.$$

So, we have

$$xd(h) = xft,$$

and thus $R(xf, xd(h))$. So, for all morphisms $f : A \to Y$, $h : A \to Y^A$ and for all $x : 1 \to A$, if $R'(f, h, x, x)$, then $R(xf, xd(h))$.

By CTGDT we have that R has a fixed point.   ▪

In [Law69] it is shown that the result that we proved is responsible for a great variety of diagonalization arguments: Russell's Paradox, Cantor's Theorem, and Tarski's (Undefinability of Truth) Theorem. We now present one of the two main results from [YM03].

Yanofsky Diagonal Theorem [YM03]. *If Y is a set and there are a set A and a function $f : A \times A \to Y$ such that all functions $g : A \to Y$ are representable by f (there is an $a \in A$ such that $g(\cdot) = f(\cdot, a)$), then all functions $\alpha : Y \to Y$ have a fixed point.*

Proof. Let us consider the category **Set** as being the category of sets. It is clear that **Set** is cartesian closed. Let us consider $\tilde{f} : A \to Y^A$ as being given by

$$(\tilde{f}(a_0))(a_1) := f(a_1, a_0).$$

By construction and under the previously introduced notations, since **Set** is cartesian closed, for every $h : A \to Y$ there is $x : 1 \to A$ such that for all $a : 1 \to A$,

$$\langle a, x\tilde{f} \rangle \epsilon = ah.$$

This means that $\tilde{f} : A \to Y^A$ is a weakly point-surjective morphism. By Lawvere Theorem we have that Y has the fixed point property. Hence, all functions $\alpha : Y \to Y$ have a fixed point.

∎

The previous result is confirmed in [YM03] to be responsible for: Diagonalization Lemma (that was presented in Chapter 3), Gödel's First Incompleteness Theorem, Tarski's Theorem, Parikh Sentences, Löb's Paradox, Recursion Theorem, and Von Neumann's Self-reproducing Machines.

The following result—the other main result from [YM03]—is the contrapositive of Yanofsky Diagonal Theorem.

Yanofsky Cantor's Theorem [YM03]. *If Y is a set and there is a function $\alpha : Y \to Y$ without fixed points (for all $y \in Y$, $\alpha(y) \neq y$), then for all sets A and for all functions $f : A \times A \to Y$ there is a function $g : A \to Y$ that is not representable by f, i.e., such that for all $a \in A$, $g(\cdot) \neq f(\cdot, a)$.*

The Yanofsky Cantor's Theorem was analysed in great detail in [YM03], where the author concluded that it was responsible for:

Cantor's $\mathbb{N} \subsetneq \wp(\mathbb{N})$ Theorem, Russell's Paradox, Grelling's Paradox, Liar Paradox, Richard's Paradox, Turing's Halting Problem, a Non-Recursively Enumerable Language, and an Oracle B such that $P^B \neq NP^B$.

7.2 Knaster-Tarski Theorem

In the present section, we are going to use TR to prove the Knaster-Tarski Theorem. After that, we are going to prove the Banach Mapping Theorem using directly the Knaster-Tarski Theorem, and we are going to prove the Schröder-Bernstein Theorem using, by its turn, directly the Banach Mapping Theorem. This means that we are going to trace back those three Theorems to TR.

Knaster-Tarski Theorem [Sha16, p. 9]. *If $\mathcal{X}$ is a set and $\Phi : \wp(\mathcal{X}) \to \wp(\mathcal{X})$ is a mapping that preserves set-containment, then Φ has a fixed point.*

Proof. Here we are going to apply TR by taking N as being V (the Universe of Set Theory, see, for instance, [TZ71, p. 19]). Let us consider the relation $R(X, Y)$ given by

$$\Phi(X) = Y,$$

the relation $R'(X, Y)$ given by

$$(\forall x \in Y.\ x \subseteq \Phi(x)) \wedge (\cup Y \subseteq X) \wedge (\cup Y \in Y) \wedge (\Phi(X) \subseteq \cup Y) \wedge X \in \wp(\mathcal{X}).$$

Furthermore, let us the function d given by:

$$d(X) := \cup X.$$

Let us see that the conditions of TR are satisfied for $R(X, Y)$, for $R'(X, Y)$, and for d:

(R1) Consider $\mathcal{E} = \{E \in \wp(X) \mid E \subseteq \Phi(E)\}$. We have that $\mathcal{E} \in \wp(\wp(X))$. Let us see that $R'(d(\mathcal{E}), \mathcal{E})$ is the case. By definition, we have that

$$\forall x \in \mathcal{E}.\ x \subseteq \Phi(x),$$

and

$$\cup \mathcal{E} \subseteq \cup \mathcal{E} = d(\mathcal{E}).$$

Let $E \in \mathcal{E}$. So, $E \subseteq \cup \mathcal{E}$, hence, $\Phi(E) \subseteq \Phi(\cup \mathcal{E})$. As $E \in \mathcal{E}$, we also have that $E \subseteq \Phi(E)$, therefore $E \subseteq \Phi(\cup \mathcal{E})$. Consequently,

$$\forall E \in \mathcal{E}.\ E \subseteq \Phi(\cup \mathcal{E}),$$

from where it follows that $\cup \mathcal{E} \subseteq \Phi(\cup \mathcal{E})$ and, as we also have that $\mathcal{E} \in \wp(\wp(X))$, we have that $\cup \mathcal{E} \in \wp(X)$ and $\cup \mathcal{E} \subseteq \Phi(\cup \mathcal{E})$, which implies that $\cup \mathcal{E} \in \mathcal{E}$.

We concluded that $\cup \mathcal{E} \subseteq \Phi(\cup \mathcal{E})$, so we also have that $\Phi(\cup \mathcal{E}) \subseteq \Phi(\Phi(\cup \mathcal{E}))$, from where we conclude that $\Phi(\cup \mathcal{E}) \in \mathcal{E}$, and so $\Phi(d(\mathcal{E})) \subseteq \cup \mathcal{E}$.

In all, we have

$$(\forall x \in \mathcal{E}.\ x \subseteq \Phi(x)) \wedge (\cup \mathcal{E} \subseteq d(\mathcal{E})) \wedge$$

$$(\cup \mathcal{E} \in \mathcal{E}) \wedge (\Phi(d(\mathcal{E})) \subseteq \cup \mathcal{E}) \wedge d(\mathcal{E}) \in \wp(\mathcal{X}),$$

so $R'(d(\mathcal{E}), \mathcal{E})$ is the case.

(R2) Let us consider $X, Y \in V$. Let us suppose that $R'(X, Y)$ is the case. So,

$$(\forall x \in Y.\ x \subseteq \Phi(x)) \wedge (\cup Y \subseteq X) \wedge (\cup Y \in Y) \wedge (\Phi(X) \subseteq \cup Y) \wedge X \in \wp(\mathcal{X}).$$

We have, by hypothesis, that $\Phi(X) \subseteq \cup Y$. As $\cup Y \in Y$ and $\forall x \in Y.\ x \subseteq \Phi(x)$, we have that $\cup Y \subseteq \Phi(\cup Y)$. But, as $\cup Y \subseteq X \subseteq \mathcal{X}$, we also have that $\Phi(\cup Y) \subseteq \Phi(X)$, and so $\cup Y \subseteq \Phi(X)$. In all, $\Phi(X) \subseteq \cup Y$ and $\cup Y \subseteq \Phi(X)$, hence $\Phi(X) = \cup Y = d(Y)$, i.e., $R(X, d(Y))$ holds. So, $R'(X, Y)$ implies $R(X, d(Y))$.

By TR, we conclude that $R(X, Y)$ has a fixed point, so Φ has a fixed point.

∎

Now, we show how the Banach Mapping Theorem is a direct consequence of the Knaster-Tarski Theorem.

Banach Mapping Theorem [Sha16, p. 8]. *Given sets X and Y, and functions $f : X \to Y$ and $g : Y \to X$, there is a subset A of X whose complement is the g-image of the complement of $f(A)$.*

Proof. Let us consider $\Phi : \wp(X) \to \wp(X)$ given by

$$\Phi(E) := X \setminus g(Y \setminus f(E)).$$

It is clear that Φ is a mapping that preserves set-containment, so, by the Knaster-Tarski Theorem, there is $A \in \wp(X)$ such that $A = \Phi(A)$. So,

$$X \setminus A = X \setminus \Phi(A) = X \setminus (X \setminus g(Y \setminus f(A))) = g(Y \setminus f(A)).$$

■

Finally, we show that the Schröder-Bernstein Theorem is a corollary of the Banach Mapping Theorem.

Schröder-Bernstein Theorem [Sha16, p. 8]. *If X and Y are sets for which there is a one-to-one mapping taking X into Y and a one-to-one mapping taking Y into X, then there is a one-to-one mapping taking X onto Y.*

Proof. Let us suppose that $f : X \to Y$ and $g : Y \to X$ are one-to-one. By the Banach Mapping Theorem, there is $A \in \wp(X)$ such that $X \setminus A = g(Y \setminus f(A))$. Let us define $h : X \to Y$ given by:

$$h(x) = \begin{cases} f(x), & x \in A \\ g^{-1}(x), & x \in X \setminus A \end{cases}$$

It is clear that h satisfies the desired conditions.

■

7.3 General Fixed Point Theorem

In this section we are going to formalise a broad notion of limit and state some conditions for a fixed point theorem for that broad limits. After that, we will analyse two fixed point theorems—one from Set Theory and the other from Functional Analysis.

The following definition allows to consider some important sequences.

Definition 13. Given $X \neq \emptyset$, we say that *a function* $f : X \to X$ *is preserved in* $\mathcal{A} \subseteq X^{\mathbb{N}}$ (where $X^{\mathbb{N}}$ denotes the class of all functions (sequences) from $\mathbb{N}$ to X) if[1], for each $x \in \mathcal{A}$, $\langle f(x_n)\rangle_{n\in\mathbb{N}} \in \mathcal{A}$.

The next definition generalises some important properties of the notion of limit.

Definition 14. Let $X \neq \emptyset$ and $\mathcal{A} \subseteq X^{\mathbb{N}}$. We say that *a function* $F : \mathcal{A} \to X$ *is a limit function* if:

Lim1.) For each $x \in \mathcal{A}$, $\langle x_{n+1}\rangle_{n\in\mathbb{N}} \in \mathcal{A}$;

Lim2.) For each $x \in \mathcal{A}$, $F(x) = F(\langle x_{n+1}\rangle_{n\in\mathbb{N}})$.

The following definition captures the notion of continuity in this broad sense.

Definition 15. Given $X \neq \emptyset$, we say that *a function* $f : X \to X$ *is continuous with respect to a function* $F : \mathcal{A} \to X$ if f is preserved in $\mathcal{A}$, if F is a limit function, and if the following conditions are satisfied:

Cont1.) There is an $\alpha \in X$ such that $\langle f^n(\alpha)\rangle_{n\in\mathbb{N}} \in \mathcal{A}$;

Cont2.) For each $x \in \mathcal{A}$, $F(\langle f(x_n)\rangle_{n\in\mathbb{N}}) = f(F(x))$.

Now the general theorem that garanties the existence of fixed points for this broad notion of limit:

---

[1]Given $x \in X^{\mathbb{N}}$ (a sequence), we will use the notation x_n, for each $n \in \mathbb{N}$, to represent the n-th element of the sequence, that is, $x(n)$.

General Fixed Point Theorem. *Let $X \neq \emptyset$, $A \subseteq X^{\mathbb{N}}$, and f be a continuous function with respect to $F : A \to X$. We have that f has a fixed point.*

Proof. Let us consider the relations $R(x,y)$ and $R'(x,y)$ in A given by, respectively, $f(F(x)) = F(y)$ and $f(F(x)) = F(\langle f(y_n) \rangle_{n \in \mathbb{N}})$. Let us take $d : A \to A$ given, for each $x \in A$, by $d(x) = \langle f(x_n) \rangle_{n \in \mathbb{N}}$.

Let us suppose that $R'(x,y)$ is the case. Then, $f(F(x)) = F(\langle f(y_n) \rangle_{n \in \mathbb{N}})$, and so $f(F(x)) = F(d(y))$, i.e., $R(x,d(y))$. Hence, $R'(x,y)$ implies $R(x, d(y))$.

Take $\alpha \in X$ in the conditions of Cont1.). Consider $z \in A$ given, for each $n \in \mathbb{N}$, by $z_n = f^n(\alpha)$. By Lim1.), Lim2.), and Cont2.), we conclude that

$$f(F(d(z))) = f(F(\langle f(z_n) \rangle_{n \in \mathbb{N}})) = F(\langle f(f(z_n)) \rangle_{n \in \mathbb{N}}) = F(\langle f^{n+2}(\alpha) \rangle_{n \in \mathbb{N}}) =$$

$$= F(\langle f^{(n+1)+1}(\alpha) \rangle_{n \in \mathbb{N}}) = F(\langle f^{n+1}(\alpha) \rangle_{n \in \mathbb{N}}) = F(\langle f(z_n) \rangle_{n \in \mathbb{N}}).$$

Hence, $R'(d(z), z)$.

Consequently, by TR we conclude that $R(x,y)$ has, at least, one fixed point, say $h \in A$. So, $R(h,h)$, that is, $f(F(h)) = F(h)$. Therefore, f has, at least, one fixed point.

7.3.1 Fixed point Lemma for Normal Functions

We are going to use the previous theorem to prove the Fixed Point Lemma for Normal Functions.

Definition 16. We say that $f :$ On $\to$ On is a *normal function* if f satisfies the following conditions:

(1) $\alpha < \beta \to f(\alpha) < f(\beta)$;

(2) If λ is a limit ordinal, then $f(\lambda) = \sup\{f(\alpha) \mid \alpha < \lambda\}$.

In [Dev94, p. 73] it is proved that if f is a normal function and $\alpha \in \text{On}$, then $\alpha \leq f(\alpha)$, and in [Lev79, p. 117] it is proved that, given a non-empty set of ordinals A and a normal function f, we have

$$f(\sup A) = \sup\{f(\alpha) \mid \alpha \in A\}. \tag{Sup}$$

Fixed Point Lemma for Normal Functions [Veb08],[Dev94, p. 73].
Every normal function has a fixed point.

Proof. Consider f a normal function. Let us take the class $\mathcal{A} = \{x \in \text{On}^{\mathbb{N}} \mid x \text{ is increasing}\}$. Given $x \in \mathcal{A}$, let us take $\overline{x} = \{x_n \mid n \in \mathbb{N}\}$. As f is normal, by condition (1) of the definition, given $x \in \mathcal{A}$, $\langle f(x_n)\rangle_{n\in\mathbb{N}} \in \mathcal{A}$; consequently f is preserved in $\mathcal{A}$.

Let us see that the conditions of the General Fixed Point Theorem are satisfied. It is clear that $F : \mathcal{A} \to \text{On}$ given, for each $x \in \mathcal{A}$, by $F(x) = \sup \overline{x}$ is a limit function. Let us consider $\beta \in \text{On}$ and $\alpha = \langle \alpha_n \rangle_{n\in\mathbb{N}}$ given by:

$$\begin{cases} \alpha_0 = \beta \\ \alpha_{n+1} = f(\alpha_n), \quad n \in \mathbb{N} \end{cases}$$

It is clear, by what was observed, that α is increasing, that is, $\alpha \in \mathcal{A}$. Let $x \in \mathcal{A}$. So, as x is increasing, we have

$$F(\langle f(x_n)\rangle_{n\in\mathbb{N}}) = \sup \overline{\langle f(x_n)\rangle_{n\in\mathbb{N}}} = \sup\{f(x_n) \mid n \in \mathbb{N}\} =$$
$$f(\sup\{x_n \mid n \in \mathbb{N}\}) = f(\sup \overline{x}) = f(F(x)).$$

In all, f is continuous with respect to F, so, by the General Fixed Point Theorem, f has, at least, one fixed point.

∎

The previous theorem was proved using the General Fixed Point Theorem and, this last result, in its turn, depends on TR. So, the Fixed Point Lemma for Normal Functions also follows from TR by considering, in $\mathcal{A}$, $R(x,y)$ as being $f(\sup \overline{x}) = \sup \overline{y}$, $R'(x,y)$ as being $f(\sup \overline{x}) = \sup(\langle f(y_n)\rangle_{n\in\mathbb{N}})$, and the function $d : \mathcal{A} \to \mathcal{A}$ such that, for each $x \in \mathcal{A}$, $d(x) = \langle f(x_n)\rangle_{n\in\mathbb{N}}$.

In [Mos06, p. 76] it is presented the Continuous Least Fixed Point
Theorem. This theorem is also a consequence of the General Fixed
Point Theorem and the proof is very similar to the previous proof—instead of considering the order in the ordinals, one considers a general
inductive order and countably continuous mappings (they are mappings which satisfy the (Sup) property).

7.3.2 Banach Fixed Point Theorem

In order to state the Banach Fixed Point Theorem we need the
following definitions (the reader can easily find them in books on
Functional Analysis or Topology, for instance, we considered [Con14],
[SV06], and [KK01]).

Definition 17. We say that $\langle X, d \rangle$ is a metric space if d is a function
from $X \times X$ to[2] $\mathbb{R}_0^+$ and if, for each $x, y, z \in X$, the following properties
are verified:

(1) $d(x,y) = 0 \leftrightarrow x = y$;

(2) $d(x,y) = d(y,x)$ (symmetry);

(3) $d(x,z) \leq d(x,y) + d(y,z)$ (triangular inequality).

Definition 18. Let $\langle X, d \rangle$ be a metric space and $x \in X^{\mathbb{N}}$. We say that
x is a *Cauchy sequence* if

$$\forall \varepsilon > 0.\ \exists N \in \mathbb{N}.\ \forall m, n > N.\ d(x_m, x_n) < \varepsilon.$$

We say that *x is convergent* if

$$\exists L \in X.\ \forall \varepsilon > 0.\ \exists N \in \mathbb{N}.\ \forall n > N.\ d(x_n, L) < \varepsilon.$$

It is a well-known fact that the limit of a convergent sequence $x \in$
$X^{\mathbb{N}}$ is unique. We will use $\lim x_n$ or $\lim x$ to denote the limit of a
convergent sequence $x \in X^{\mathbb{N}}$.

---

[2]Here $\mathbb{R}_0^+ = \{x \in \mathbb{R} \mid x \geq 0\}$.

Definition 19. Let $\langle X, d \rangle$ be a metric space. We say that $\langle X, d \rangle$ is *complete* if every Cauchy sequence is convergent.

Definition 20. Let $\langle X, d \rangle$ be a metric space. A function $T : X \to X$ is called a *contraction mapping (on X)* if

$$\exists q \in [0,1). \; \forall x, y \in X. \; d(T(x), T(y)) \leq qd(x,y).$$

We are going to prove Banach Fixed Point Theorem using the General Fixed Point Theorem.[3]

Banach Fixed Point Theorem [KK01]. *Let $\langle X, d \rangle$ be a non-empty complete metric space with a contraction mapping $T : X \to X$. Then, T has a fixed point.*

Proof. Let $\mathcal{A} = \{x \in X^{\mathbb{N}} \mid x \text{ is a Cauchy sequence}\}$. Take $q \in [0,1)$ such that

$$\forall x, y \in X. \; d(T(x), T(y)) \leq qd(x,y).$$

Consider $x \in \mathcal{A}$. So, as x is a Cauchy sequence,

$$\forall \varepsilon > 0. \; \exists N \in \mathbb{N}. \; \forall m, n > N. \; d(x_m, x_n) < \varepsilon.$$

Let us prove that $\langle T(x_n) \rangle_{n \in \mathbb{N}} \in \mathcal{A}$, that is, let us prove that $\langle T(x_n) \rangle_{n \in \mathbb{N}}$ is a Cauchy sequence. Let $\varepsilon > 0$. We have that there is $N \in \mathbb{N}$ such that $\forall m, n > N. \; d(x_m, x_n) < \frac{\varepsilon}{q}$. Let $m, n > N$. Hence, $d(T(x_m), T(x_n)) \leq qd(x_m, x_n) < q \cdot \frac{\varepsilon}{q} = \varepsilon$. So, $\langle T(x_n) \rangle_{n \in \mathbb{N}} \in \mathcal{A}$, as wanted. This means that T is preserved in $\mathcal{A}$.

Take $F : \mathcal{A} \to X$ given, for each $x \in \mathcal{A}$, by $F(x) = \lim x$. As $\langle X, d \rangle$ is complete, we have that F is well-defined. Furthermore, it is obvious that F is a limit function.

Let us prove that T is continuous with respect to F. Let $x_0 \in X$. Consider $\alpha \in X^{\mathbb{N}}$ given, for each $n \in \mathbb{N}$, by $\alpha_n = T^n(x_0)$. Let us prove that $\alpha \in \mathcal{A}$. For that, let us firstly prove, by induction on n, that:

$$\forall n \in \mathbb{N}. \; d(\alpha_{n+1}, \alpha_n) \leq q^n d(\alpha_1, \alpha_0). \tag{I}$$

---

[3]We will only state the existence part of the Banach Fixed Point Theorem, because the unicity part does not use a diagonalization reasoning.

For $n = 0$ the result is obvious. Let us suppose, by induction hypothesis, that (I) hold for $n \in \mathbb{N}$. Then,

$$d(\alpha_{(n+1)+1}, \alpha_{n+1}) = d(\alpha_{n+2}, \alpha_{n+1}) = d(T(\alpha_{n+1}), T(\alpha_n)) \leq q d(\alpha_{n+1}, \alpha_n)$$
$$\leq q^{n+1} d(\alpha_1, \alpha_0).$$

So, (I) holds. Let us now prove that α is a Cauchy sequence. Let $\varepsilon > 0$. Take $N \in \mathbb{N}$ such that

$$q^N < \frac{\varepsilon(1-q)}{d(\alpha_1, \alpha_0)}.$$

Let $m, n > N$ be such that $m > n$. Then, by the triangular inequality, by the fact that $\sum_{k=0}^{\infty} q^k = \frac{1}{1-q}$ as $q \in [0, 1)$, and by (I), we have that

$$d(\alpha_m, \alpha_n) \leq \sum_{i=n}^{m-1} d(\alpha_{i+1}, \alpha_i) \leq \sum_{i=n}^{m-1} q^i d(\alpha_1, \alpha_0) = q^n d(\alpha_1, \alpha_0) \sum_{k=0}^{m-n-1} q^k$$

$$\leq q^n d(\alpha_1, \alpha_0) \sum_{k=0}^{\infty} q^k = q^n d(\alpha_1, \alpha_0) \left(\frac{1}{1-q} \right)$$

$$< q^N d(\alpha_1, \alpha_0) \left(\frac{1}{1-q} \right) < \frac{\varepsilon(1-q)}{d(\alpha_1, \alpha_0)} d(\alpha_1, \alpha_0) \left(\frac{1}{1-q} \right) = \varepsilon.$$

Consequently, α is a Cauchy sequence, i.e., $\alpha \in \mathcal{A}$. Let $x \in \mathcal{A}$. As every contraction mapping is continuous (in the usual topological sense), we have that

$$F(\langle T(x_n) \rangle_{n \in \mathbb{N}}) = \lim T(x_n) = T(\lim x_n) = T(\lim x) = T(F(x)).$$

In all, T is continuous with respect to F. So, by the General Fixed Point Theorem, we conclude that T has a fixed point.   ∎

Once again, the previous theorem was proved using the General Fixed Point Theorem that depends on TR. So, the Banach Fixed Point Theorem also follows from (TR) by considering, in $\mathcal{A}$, $R(x, y)$ as being $T(\lim x) = \lim y$, $R'(x, y)$ as being $T(\lim x) = \lim \langle f(y_n) \rangle_{n \in \mathbb{N}}$, and the function $d : \mathcal{A} \to \mathcal{A}$ such that, for each $x \in \mathcal{A}$, $d(x) = \langle T(x_n) \rangle_{n \in \mathbb{N}}$.

In [Sha16, p. 31–36] it is shown that Newton's Method and the Initial-Value Problem follow from the Banach Fixed Point Theorem.

8 Conclusions and Future Work

8.1 Conclusions

The main objective of the present work was to study diagonalization in a formal system with a view towards a general theory of diagonalization that can be applied to everyday Mathematics. We started to study in detail the Diagonalization Lemma in Chapter 3, then we moved to argue that Yablo's Paradox is self-referential in Chapter 4. After that, in Chapter 5, we presented a common origin of several paradoxes and Löb's Theorem; furthermore, we presented a general approach to paradoxes. In Chapter 6 we presented a general theory of diagonalization. Finally, several Mathematical examples were studied in Chapter 7 using the theory presented in Chapter 6.

The Diagonalization Lemma, as studied in Chapter 3, gives rise to self-referential sentences, a concept that was generalised to the notion of a general formula. We argued that very natural properties related to self-reference are not decidable in a theory of Arithmetic (a consistent primitively recursive axiomatised extension of PA). For example, if $\sigma(x)$ is a predicate that identifies all self-referential formulas in T (a theory of Arithmetic), then $\sigma(x) \wedge \mathcal{P}_T(x)$ is not decidable: using theory T we cannot decide which sentences are both self-referential and provable (see Theorem 4). This might seem to be an expected and intuitive result when one has in mind that in a theory of Arithmetic there are always incompleteness phenomena, nevertheless it is always important to provide an argument or a proof to such claims—just like was done in Chapter 3.

Besides the Diagonalization Lemma—that states the diagonalization of a formula and that is deeply related with self-reference—, in Chapter 3 we also presented the Strong Diagonalization Lemma—that

© Springer Fachmedien Wiesbaden GmbH, part of Springer Nature 2020
P. G. Santos, *Diagonalization in Formal Mathematics*, BestMasters,
https://doi.org/10.1007/978-3-658-29111-2_8

is similar to the former result but states a diagonalization at the level of terms. It is a well-known fact that from the Strong Diagonalization Lemma one can prove the Diagonalization Lemma. Despite that fact, we were able to prove that the conserve does not hold—Corollary 2—by firstly proving that the Diagonalization Lemma cannot prove itself—Theorem 5. This last result is an explicit example of a situation where the Logicism program fails: the diagonalization of terms—formal Arithmetic—cannot be reduced to the diagonalization of formulas—pure Logic.

In Chapter 4 we presented Yablo's Paradox and a minimal theory that does not require any arithmetical fact to express that contradiction, theory $\mathcal{Y}$ (see Definition 3, Theorem 6, and Theorem 7). Inspired by theory $\mathcal{Y}$, we introduced Linear Temporal Logic (LTL) and the main concepts needed to study Yablo's Paradox. From that, we argued that a formulation of Yablo's Paradox in LTL—(TY)—was sound and captured the main features of the paradox. Finally, we argued that our formulation of Yablo's Paradox in LTL is self-referential.

Paradoxes and Löb's Theorem were the main focus of Chapter 5. We started by presenting a result by Smullyan—Theorem R—that was confirmed to be responsible for all the diagonalization phenomena presented in the considered chapter. After that, we exhibit several paradoxes (the Liar, Russell's Paradox, and Curry's Paradox) and also Löb's Theorem. From that point, all the technical apparatus was introduced as well as the systems that generalise the considered paradoxes and theorem: Curry Systems (see Definition 7). Not only the layout of Curry's Paradox was preserved in Curry Systems, but also the fact that one very specific formula could be deduced was preserved—Theorem 5.4.

In order to show that Löb's Theorem is a particular case of a Curry System, the notion of Löb System (Definition 8) was introduced. To fulfil the mentioned goal, we proved that every Löb System is a Curry System—Theorem 10—and that Löb's Theorem holds in every Löb System—Löb Theorem. Then, a general system to study paradoxes was introduced—Definition 9. To show that in some situations consistency was preserved, it was introduced a notion of model of a Curry System (and of a Löb System)—Definition 10—and a consistency result was proved—Consistency Theorem.

In Chapter 6 was introduced a general theory of diagonalization—diagonal theory (see Definition 12). It was proved that in a diagonal theory one of the formulas can be diagonalized—General Diagonalization Theorem. We ended Chapter 6 with several structural results (in the sense that they are not in a formal language but rather in a everyday Mathematics fashion) that follow from the General Diagonalization Theorem and that were used to study several Mathematical examples.

Finally, in Chapter 7 several examples from Mathematics were studied and traced back to the General Diagonalization Theorem: 1.1.Theorem from [Law69], two results from [YM03], the Knaster-Tarski Theorem, the Banach Mapping Theorem, the Schröder-Bernstein Theorem, and a General Fixed Point Theorem responsible for the Fixed Point Lemma for Normal Functions and for the Banach Fixed Point Theorem (that by its turn is responsible for Newton's Method and the Initial-Value Problem).

8.2 Future Work

Several lines of investigation could be followed having as a starting point the current thesis. Regarding self-reference and what was developed in Chapter 3, one very natural question is "What is the correct way to formulate self-reference?"; a model-theoretic approach could be considered or even different formal systems. It can be the case that partial notions of self-reference could be totally studied in specific formal systems. Another interesting topic of investigation is the study of other paradoxes using Definition 9. Finally, other examples of diagonalization in Mathematics could be studied (for example results from Topology that were not considered in the present thesis due to their high complexity to be clearly stated).

Bibliography

[Alo05] Maria Aloni. Individual concepts in modal predicate logic. *Journal of Philosophical Logic*, 34(1):1–64, 2005.

[AM99] R.J.L. Adams and R. Murawski. *Recursive Functions and Metamathematics: Problems of Completeness and Decidability, Gödel's Theorems*. Synthese Library. Springer, 1999.

[Bar93] John Barwise. *Handbook of Mathematical Logic*. North-Holland, eighth edition, 1993.

[Bea13] Jc. Beall. Curry's paradox. In Edward N. Zalta, editor, *The Stanford Encyclopedia of Philosophy*. Metaphysics Research Lab, Stanford University, spring 2013 edition, 2013.

[BGR16] Jc. Beall, Michael Glanzberg, and David Ripley. Liar paradox. In Edward N. Zalta, editor, *The Stanford Encyclopedia of Philosophy*. Metaphysics Research Lab, Stanford University, winter 2016 edition, 2016.

[Boo84] George Boolos. *The logic of provability*, volume 91. Cambridge university Press, 1984.

[Bow79] Kenneth A. Bowen. *Model Theory for Modal Logic*. Springer-Science+Business Media, B.V., 1979.

[Bus98] Samuel R. Buss. *Handbook of Proof Theory*, volume 137. Elsevier, 1998.

[Con14] John B. Conway. *A Course in Point Set Topology*. Undergraduate Texts in Mathematics. Springer International Publishing, 1 edition, 2014.

© Springer Fachmedien Wiesbaden GmbH, part of Springer Nature 2020
P. G. Santos, *Diagonalization in Formal Mathematics*, BestMasters,
https://doi.org/10.1007/978-3-658-29111-2

[Coo14] R.T. Cook. *The Yablo Paradox: An Essay on Circularity*. OUP
         Oxford, 2014.

[Dev94] Keith Devlin. *The Joy of Sets. Fundamentals of Contemporary
         Set Theory*. Undergraduate Texts in Mathematics. Springer,
         2nd edition, 1994.

[EFT96] H.-D. Ebbinghaus, J. Flum, and W. Thomas. *Mathematical
         Logic*. Springer, second edition, 1996.

[ES08]   Peter Eldridge-Smith. *The Liar Paradox and its Relatives*.
         PhD thesis, The Australian National University and
         Philosophy Program, School of Humanities, 3 2008.

[GG15]   Valentin Goranko and Antony Galton. Temporal logic.
         In Edward N. Zalta, editor, *The Stanford Encyclopedia of
         Philosophy*. Metaphysics Research Lab, Stanford University,
         winter 2015 edition, 2015.

[HV14]   Volker Halbach and Albert Visser.   Self-reference in
         arithmetic i. *The Review of Symbolic Logic*, 7(4):671–691,
         2014.

[HZ17]   Volker Halbach and Shuoying Zhang. Yablo without gödel.
         Analysis, 77(1):53–59, 2017.

[ID16]   Andrew David Irvine and Harry Deutsch. Russell's paradox.
         In Edward N. Zalta, editor, *The Stanford Encyclopedia of
         Philosophy*. Metaphysics Research Lab, Stanford University,
         winter 2016 edition, 2016.

[Jer73] R. G. Jeroslow.   Redundancies in the hilbert-bernays
         derivability conditions for gödel's second incompleteness
         theorem. *J. Symbolic Logic*, 38(3):359–367, 12 1973.

[Ket04] Jeffrey Ketland. Bueno and colyvan on yablo's paradox.
         Analysis, 64(2):165–172, 2004.

[KK01] Mohamed A. Khamsi and William A. Kirk. *An Introduction to Metric Spaces and Fixed Point Theory*. Wiley-Interscience, 1 edition, 2001.

[KM08] Fred Kroger and Stephan Merz. *Temporal Logic and State Systems*. Texts in Theoretical Computer Science. An EATCS Series. Springer, 2008.

[Lan13] S.M. Lane. *Categories for the Working Mathematician*. Graduate Texts in Mathematics. Springer New York, 2013.

[Law69] F. William Lawvere. Diagonal arguments and cartesian closed categories. In *Category Theory, Homology Theory and their Applications II*, pages 134–145, Berlin, Heidelberg, 1969. Springer Berlin Heidelberg.

[Lev79] Azriel Levy. *Basic Set Theory*. Springer-Verlag Berlin Heidelberg GmbH, 1979.

[Lin06] Per Lindström. Note on some fixed point constructions in provability logic. *Journal of Philosophical Logic*, 35(3):225–230, 2006.

[Lin17] P. Lindström. *Aspects of Incompleteness*. Lecture Notes in Logic. Cambridge University Press, 2017.

[Löb55] M.H. Löb. Solution of a problem of leon henkin. *The Journal of Symbolic Logic*, 24:115–118, 1955.

[Mos06] Yiannis Moschovakis. *Notes on Set Theory*. Springer, segunda edition, 2006.

[Pic13] Lavinia María Picollo. Yablo's paradox in second-order languages: Consistency and unsatisfiability. *Studia Logica*, 101(3):601–617, Jun 2013.

[Pic18] Lavinia Picollo. Reference in arithmetic. *The Review of Symbolic Logic*, 11(3):573–603, 2018.

[Pri97]   Graham Priest. Yablo's paradox. *Analysis*, 57(4):236–242, 1997.

[Raa18]   Panu Raatikainen. Gödel's incompleteness theorems. In Edward N. Zalta, editor, *The Stanford Encyclopedia of Philosophy*. Metaphysics Research Lab, Stanford University, fall 2018 edition, 2018.

[Rau06]   W. Rautenberg. *A Concise Introduction to Mathematical Logic*. Universitext (Berlin. Print). Springer, 2006.

[Sha16]   Joel H. Shapiro. *A Fixed- Point Farrago*. Springer, 2016.

[Sho18]   J.R. Shoenfield. *Mathematical Logic*. CRC Press, 2018.

[SK17]   Paulo G. Santos and Reinhard Kahle. Diagonalização, paradoxos e o teorema de löb. *Revista Portuguesa de Filosofia*, 73(3–4):1169–1188, 2017.

[Smi13]   Peter Smith. *An Introduction to Gödel's Theorems*. Cambridge Introductions to Philosophy. Cambridge University Press, 2 edition, 2013.

[Smu94]   Raymond M. Smullyan. *Diagonalization and Self-reference*. Oxford Science Publications, 1994.

[SV06]   Satish Shirali and Harkrishan L. Vasudeva. *Metric Spaces*. Springer-Verlag, London, 2006.

[Ten17]   Neil Tennant. Logicism and neologicism. In Edward N. Zalta, editor, *The Stanford Encyclopedia of Philosophy*. Metaphysics Research Lab, Stanford University, winter 2017 edition, 2017.

[TZ71]   G. Takeuti and W.M. Zaring. *Introduction to Axiomatic Set Theory*. Springer-Verlag BErlin Heidelberg GmbH, 1971.

[Und16]   Decidable formula. In *Encyclopedia of Mathematics*. Springer, 2016.

[Veb08] Oswald Veblen. Continuous increasing functions of finite
         and transfinite ordinals. *Transactions of the American
         Mathematical Society*, 9(3):280–292, 1908.

[Ver17] Rineke Verbrugge. Provability logic. In Edward N. Zalta,
         editor, *The Stanford Encyclopedia of Philosophy*. Metaphysics
         Research Lab, Stanford University, summer 2017 edition,
         2017.

[Yab93] Stephen Yablo. Paradox without self–reference. *Analysis*,
         53(4):251–252, 1993.

[YM03]   Noson S. Yanofsky and Saunders Maclane. A universal
         approach to self-referential paradoxes , incompleteness
         and fixed points. *ArXiv*, pages 1–24, 2003.

Printed in the United States
By Bookmasters